Neurons, Axons, Dendrites, Synapses, and Memory: My Life

To Marcia
with love
Joe Morales

Neurons, Axons, Dendrites, Synapses, and Memory: My Life

Jose Morales Dorta

Library of Congress Control Number:		2019909447
ISBN:	Hardcover	978-1-7960-4536-9
	Softcover	978-1-7960-4535-2
	eBook	978-1-7960-4534-5

Print information available on the last page.

Rev. date: 07/18/2019

To order additional copies of this book, contact:
Xlibris
1-888-795-4274
www.Xlibris.com
Orders@Xlibris.com
798088

Book Review of Dr. Jose Morales Dorta's book: Neurons, Axons, Synapses, Memory: My Life

I was surprised to find myself enjoying a science book since I am a lover of fiction. Dr. Dorta's book about the making and changing of memories is fascinating.

I knew our memories are not always 100% accurate. Every book, movie, or television program about police work shows that eye witness accounts are not reliable. I did not know, however, until I read Dr. Dorta's book, that our brain can change our memories to fit what we believe is true. I also learned that our traumatic memories can even be erased with the help of a therapist.

The book is full of scientific information about neurons, synapses, and DNA; yet expressed in a way that is easy to read and understand. Just for fun, I researched some of the facts which Dr. Dorta had presented that I thought were a bit questionable or fanciful. I verified the accuracy of each and every one.

Dr. Dorta has no qualms about sharing his life experiences, both joyful and traumatic. He used his personal experiences and those of his patients to demonstrate the power and resilience of the human brain.

Dr. Dorta stresses the need for more research on a global level to understand and, hopefully, cure some of our most daunting diseases such as Parkinson's, Alztheimer's, and other brain related diseases. These are particularly devastating diseases for the patient and their families.

I know that each reader of this book will understand more about the functions of the human brain and the potential for growth, repair, and sustainability of its power. The human brain is a unique organ of unfathomable intricacies. I applaud Dr. Dorta's success in tackling such a subject for us.

Cathy Williamson

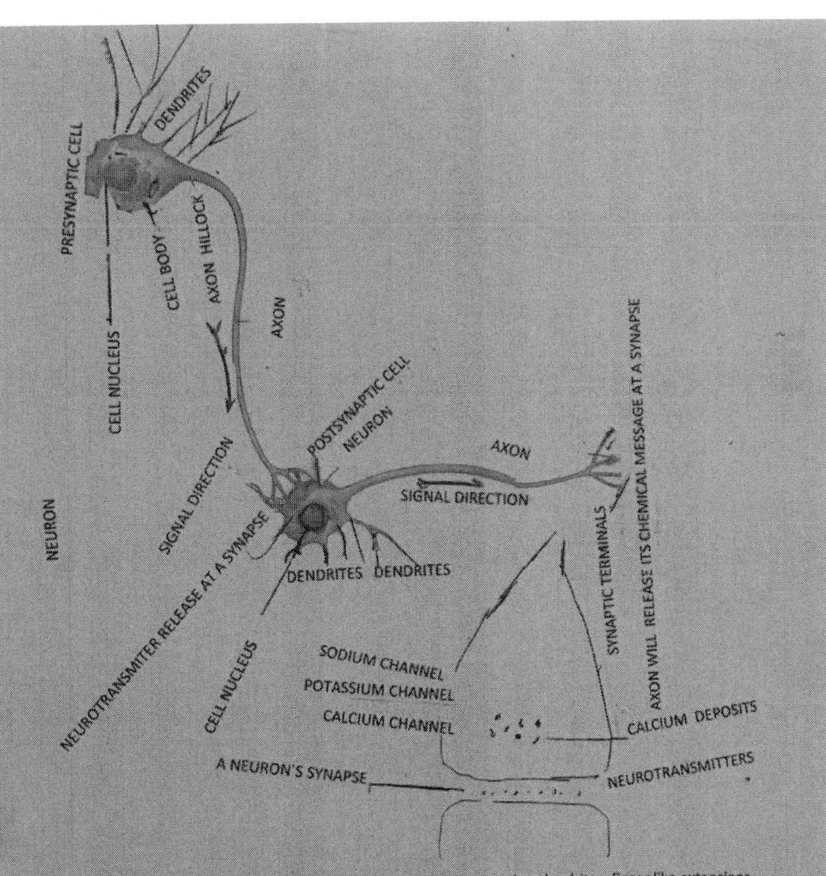

Neurons are nerve cells with component parts. Each neuron has dendrites, finger like extensions, counted by the thousands. They gather information transmitted by other nerve cells. Each neuron also has a relative long extension from the cell body called axon that transmit information to other cells. The axon carries an electrical message originated in the cell body. This electrical message is translated into a chemical message that will be released at a synapse. It goes from a presynaptic neuron to a postsynaptic neuron. Most neuro-scientists estimate that there are one hundred billion neurons in our brain. Besides neurons, there are four to six times glia cells; these are non-coding nerve cells. Some glia cells are part of our defense system, while others produce myelin to insulate the axon thus helping deliver the electrical current to its target. MIT biologists have discovered that neurons in our brain can become damaged by free radicals. This damage known as oxidative stress produces an unusual pileup of short snippets (fragments) of RNA in some neurons. They target a part of the brain called striatum. It is a particular site for Huntington and Parkinson disease. The hippocampi and amygdalae are neuronal regional part of the brain extremely sensitive to chronic stress.

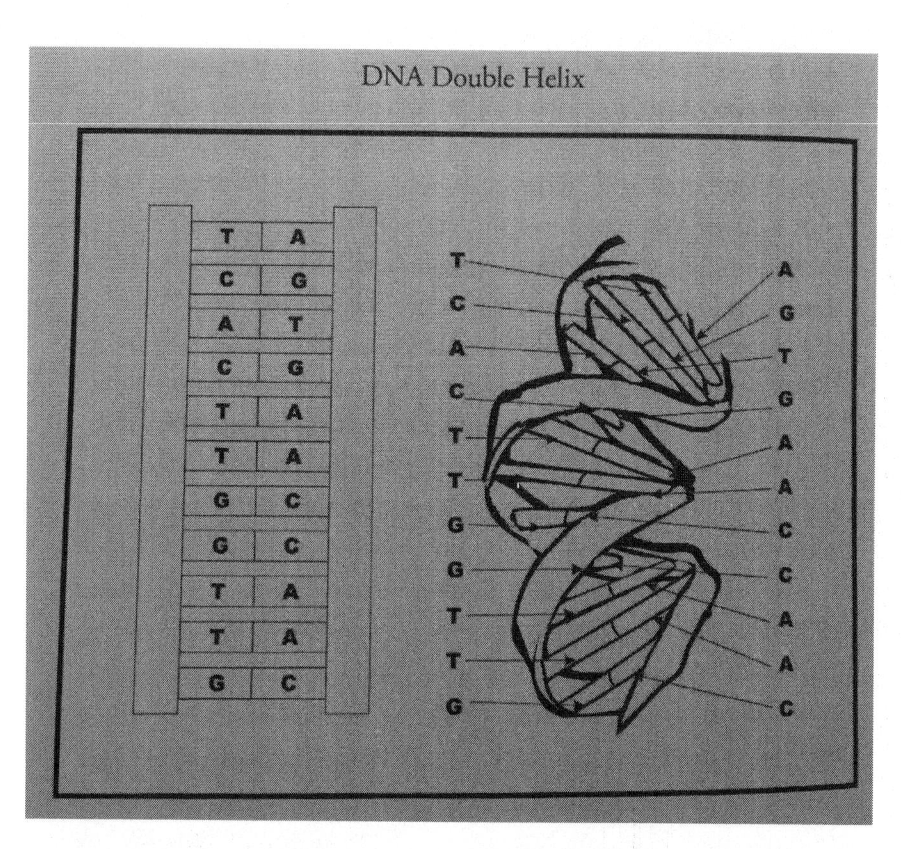
DNA Double Helix
T A
C G
A T
C G
T A
T A
G C
G C
T A
T A
G C
T
C
A
C
T
T
G
G
T
T
G
A
G
T
G
A
A
C
C
A
A
C

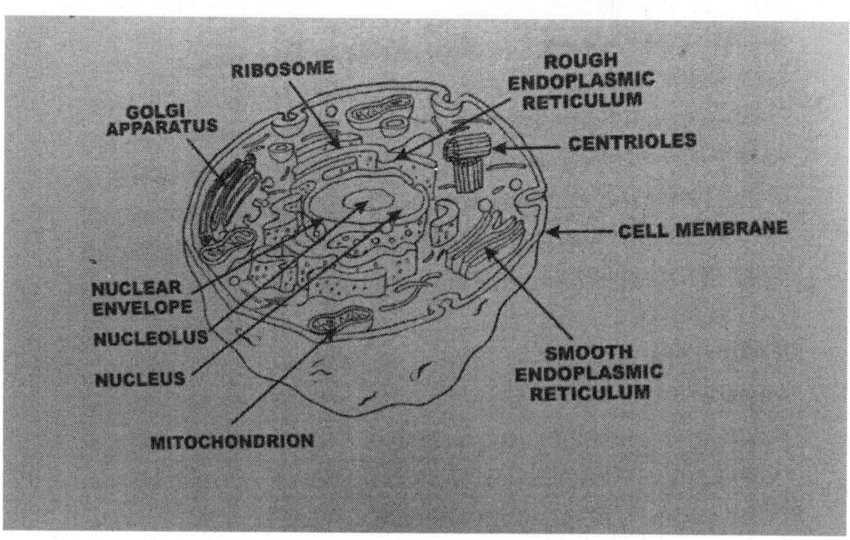
RIBOSOME
ROUGH ENDOPLASMIC RETICULUM
GOLGI APPARATUS
CENTRIOLES
CELL MEMBRANE
NUCLEAR ENVELOPE
NUCLEOLUS
NUCLEUS
SMOOTH ENDOPLASMIC RETICULUM
MITOCHONDRION

Introduction

My curiosity on memory storage and retrieval began very early during my adolescence. My attention was focused on learning how my brain connects my impressions, feelings, experiences, and speculative thoughts into permanently stored memories. Two incidents fueled this curiosity. It used to amuse me that I was reprimanded by my oldest sister for hugging and attempting to kiss a girl when I was four years old. Had I seen that demonstration of affection among members of my family and friends before?

There was another recurrent memory that often frightened me. When I was around five years old, I fell into a pond while I was playing with my dog. I could not swim. My dog jumped right after me. I grabbed him by his leg, and we both made it safely out of the water. I have always felt that I would have drowned if not for my dog. Those two memories used to amuse and frighten me equally for unknown reasons.

At the time I was not interested in my brain's dreams, and I knew nothing about unconscious memory. I thought dreaming was my soul traveling in space and had no meaningful relationship with daily life experiences. Fairy tales told by my parents at night used to confirm my belief on dreams. Fairy tales were an effective nighttime mental exercise used by parents and grandparents to keep the family together before going to bed. My parents also used it as a memory rehearsal by testing my ability to remember the main character from the previous night's story. If successful, we received special approval from my father. As a reward, they gave us a present for remembering good tales at nighttime!

Chapter I

Assembly of Bits and Pieces

At present, my curiosity is centered on how my brain selects bits of information formed during the day and joins it with past conscious and unconscious memories. Someone curious enough in memory formation may ask me, how can we find novel and interesting things in the brain if dreams are not included?

Dreams have been used by philosophers, healers, theologians, and storytellers for centuries to tell the future and fate of people. Many psychologists use dream interpretation to explore the workings of the human mind. Sigmund Freud's theory of the unconscious mind is used in many parts of the world to bring relief to anxiety disorders, depression, and many stressful situations. Freud relied upon dreams and free associations to explore and analyze past repressed memories. Repressed memories are unconsciously "forgotten" memories that have been stored in our brains. In most cases, these memories are "forgotten" because they provoke pain or anxiety.

Repressed memories do not usually appear during a conscious conversation of a normal mentally healthy person. Sometimes, however, during a normal conversation, we say something we did not consciously intended to say, and we feel embarrassed. This is known as a Freudian slip by some lay people and professionals in our society. Repressed memories, as well as unconsciously stored memories, are the by-product of a conscious brain. Most recently, scientists have been using brain scanners to explore the workings and physiology of the brain while we are in a sleep state. Several scientists have found different levels of brain activity while we are sleeping. It is generally agreed upon by most people interested in the dynamics of the brain that this wonder group of cells—the brain—never fully goes to sleep. The brain may close some sensory-neuronal circuits while the rest of the body is in a rest state.

Normally, I need a quiet and comfortable bed to be able to fall asleep. I need to turn off all lights on my sleeping room before I can go to sleep. I cannot have the radio or television set on before I am ready to hit my bed and say good night to anyone. It seems as if my conscious self is telling me to turn off all external stimuli before I turn off the internal sensory circuits. In other words, the conscious self is telling me to get the body's external sensory receptors: eyes, ears, and skin turned off to place me in a resting state. In this agreement between my conscious self and an unconscious assumption, I may be creating a habit.

A habit is something you repeat over and over again, sometimes at a conscious level and sometimes at an unconscious level. It also means this habit has been stored as a molecule somewhere in or around a synapse.

There are trillions of synapses in my brain. There are over a hundred billion brain cells known as neurons, and each neuron may have an average of around ten thousand synapses. There is enough room in a human's brain cells to store several trillion bits of memories. There are brain cells whose functions have not been clearly established yet. Their number amounts into several hundred billion. These cells are generally known as glia cells. These cells are considered noncoding genes cells.

Genes are fragments of your genome. The genome is a group of nucleotides, depicted by the letters *A*, *T*, *C*, and *G*, composing the double helix. These letters are generally known as nucleotides along with its sugar-phosphate backbone. Each letter with its backbone normally forms a twisted strand of the double helix. Nucleotides carry memories from our ancestors. It is our genetic code. There is a chemical attraction between *A* and *T*, and a corresponding chemical attraction between the letters *C* and *G*. These letters or nucleotides form the structure known as the double helix. This was presented to the public by J. D. Watson and Francis Crick in 1953. These letters or nucleotides form two long strands of double helixes known as our genome. Sections of our genome can be cut into pieces at specific letters, thus forming a gene. There are several restriction enzymes that normally can do that. Here, I will find or come across the letters DNA, which is our genetic information or great molecule. This genetic information is folded up by proteins into chromosomes.[1]

1. *Nature* 535 (July 14, 2016): 234.

Chapter II

Chromosomes

Chromosomes are threadlike structures found inside the nucleus of animals and plants. They are made of proteins that are the building blocks of our body. Proteins are assembled inside the cell by tiny vesicles called ribosomes. Inside the cell, there are many vesicles carrying out different and very important life functions. In humans, each cell normally contains twenty-three pairs of chromosomes. We inherit twenty-three pairs from the mother and twenty-three pairs from the father, making a total of forty-six chromosomes. Chromosomes are exposed to risk and threats. Here are some examples: a gene mutation, variants, copy number like in Down syndrome, which has three base pairs instead of two. Methylation is another health risk we encounter during our lifetime. In this last example, a methyl group (three hydrogen atoms bonded to a carbon atom) changes the structure of a gene; structure determines function. When a fraction of a genome is set to be cut in order to form a protein and it has an attached methyl group, it provokes a serious epigenetic problem. *Epigenetics* refers to beyond the four letters *A*, *T*, *D*, and *G*. I will deal with epigenetics later. If the above epigenetic problem is not corrected by our own body's editing system, the embryo may end up a sick child. The above examples are bits of information stored and carried out by our genes. These are biological processes carried out by most living organisms.

Viruses are a world apart; they need a host to make proteins for them. Coding genes possesses information to form proteins, but it is not the type of information-making memories that I am looking for.

Helping a person gain access to stored unconscious and toxic memories is a big challenge. There is a brain nucleus in each side or hemisphere of the brain, specifically, in the temporal lobe known as hippocampus. Generally, memories go first to the hippocampus for selectivity and further shelving in appropriate storage locations within the brain. We have memories loaded with feelings and emotions. There is a section or a system of our brain called limbic system that is central to our feelings. It is located deep in our brain. Also, located in each hemisphere or side of our brain is an old group of cells known as amygdalae. It is considered the door to our limbic system. It is like an alert brain switch when there is an external risk to our lives. The amygdalae must warn us when there is an external risk of life and death. It has direct connections with the prefrontal cortex to take appropriate action when needed. The amygdala can detect strong feelings of anger and joy in the face of people. Scanners can show us when the amygdala is active even when the subjects under study are not aware of their emotional state. The amygdala is a very important point of contact interface between visual and hearing stimulus of fear and anger.

Under normal conditions, messages coming from the neck go up to the thalamus. However, messages originating above the thalamus, meaning in the neocortex, go to the thalamus to be relayed to appropriate group of neurons and organs in our body. However, when the amygdalae, which are located in each temporal lobe, detect life-threatening signals, it bypasses normal circuity and contacts the autonomous nervous system (sympathetic) for immediate action. There is no time to think things over. The person must fight or flight unless panic takes over and the victim freezes in place. Messages move in milliseconds, going from one group of neurons in the amygdala destined to other neurons engaged in preparing the body for immediate action. After the body is out of danger and there is enough time to look back, the prefrontal cortex will take over.

When we are out danger, the frontal lobe makes an evaluation of the situation and acts based on past related experience. We should not overlook or ignore the fact that a dysfunctional amygdala or hippocampus would seriously compromise our organism's behavior; secondary to chronic stress, an injury, a disease, or addiction would seriously compromise our organism behavior.

How should I consider the amygdalae recognition of danger signals and alert the sympathetic system before anything else? Should it be a chemical reaction, past learned memory stored through the hippocampus, an episodic memory, an implicit or explicit memory?

There are other groups of cells that are considered excellent survival tools that are not related to the amygdalae. Dogs and bears smell objects from a long distance away. Eagles and falcons have eyesight for distance superior to dogs, bears, and humans.

"The prefrontal cortex is the seat of higher cognitive processes."[2] Most neuroscientists consider it the decision-making section of our brain. Part of the brain's limbic system is located within the frontal cortex; therefore, it participates in the decision-making process. When we walk on a poorly lighted street of a large city and we come across a tall and muscular man with his face half covered, our amygdalae will put us on high-alert status. In most cases, the amygdalae will override normal behavioral circuits and connect with the prefrontal cortex and the thalamus. An automatic response may come into play, engaging the hypothalamus, the pituitary, and the adrenal gland. Consequently, adrenaline will be running in our blood system, preparing our organism for an appropriate response. This physiological response of our organism is taking place at an unconscious level. It is like when your enzymes are breaking down mashed potatoes and a juicy steak you cooked in your backyard. You feel great sharing it with your family and friends, but you are not consciously aware how the enzymes worked it out. You may never know how enzymes break down proteins from the steak you just ate. Proteins are broken down so it can get to our blood system from our intestines. Most of us do not pay much attention to it and celebrate

---

2. *Nature* 535 (July 14, 2016): 8.

the dinner with a glass of wine. It is all right not to be overly concerned with everything inside and outside us; otherwise, our brain would be overwhelmed by excess information at the expense our survival.

As I mentioned earlier, the thalamus functions as a filter of stimuli from above itself in the neocortex and from the rest of our body. Our original matrix cell did a wonderful job dividing and multiplying itself to organize itself into clusters of specialized functional cells. It took time for DNA-RNA and the mitochondria molecules to join and become a very complex human being.

However, there is a decision to be made regarding the tall and muscular man with his face half covered, and that decision is at a conscious level. In most cases, you fight or leave the place in a hurry. In some cases, our organism fails to respond appropriately. You may freeze in place and may unconsciously collapse and fall on the ground.

My attempt to follow the processing of a memory and to classify it has become extremely complicated. In this hypothetical case, we have engaged the central and peripheral nervous system, the blood circulatory system, the adrenaline system, and the sensory systems. Most probably, the body immune system was placed on red alert in case it needs to take appropriate action. This seems to be a normal reaction or response when we are faced with or confronted with what appears to be a life-threatening situation.

However, nothing is lost if we remember that memories are processed or formed at the synaptic level in our brain. These memories will not be stored in the heart, kidney, lungs, or liver although they may be affected by it. I must continue following the result of my original objective: tracing memory formation and classification. Facing a tall man with his face half covered may turn out to be a toxic memory. Our DSM-IV (mental health disorder classification codes from the *Diagnostic and Statistical Manual of Mental Disorders*) may categorize this toxic memory as PTSD, meaning posttraumatic stress disorder. At present time, a posttraumatic stress disorder is considered a brain disease. Some of our soldiers discharged from the Vietnam, Iraq, and Afghanistan wars are victims of this brain disorder.

Chapter III

General Patton

Not long ago I saw a Hollywood film, *Patton*. It was based on World War II when General George C. Patton was advancing his armies against Adolf Hitler's armies. A soldier was hospitalized for what appeared to be a posttraumatic stress disorder. There was not a visible physical injury or illness to justify his hospitalization. His illness was in his mind, right inside his brain. PTSD did not exist as a psychiatric diagnosis at the time. General Patton believed the soldier was a malingerer trying to avoid joining his platoon in combat. General Patton, unaware that an anxiety attack could disable a soldier, slapped the soldier on his face. The soldier incident reached Washington, and several American newspapers called for General Patton to be relieved from his command. However, Patton was considered by some of his colleagues, including General Dwight Eisenhower, the best field commander the Allies had, and his commend was saved.

I have begun to characterize and name memories. I will begin with my own childhood experience. I called it a toxic memory; the psychiatric diagnostic manual considers it a mental disorder, PTSD. This disease of the brain requires psychiatric attention and care. In most cases, it is followed up with medication and psychotherapy. If not adequately treated at a professional level, a posttraumatic stress disorder may lead to a psychotic episode ending in homicide or suicide. A person suffering from a posttraumatic stress disorder may be a homeless individual lying down on the street using or abusing illegal drugs to calm down his nerves. While some PTSD victims may find relief taking heroin, cocaine, or alcohol, other victims discharge their anger and frustration on family members or on society at large.

PTSD may have its origin at home by a family member or even at school bullying by a fellow student. A boy or girl may have been abused by the parent or family member who could have, in turn, been mistreated by his or her own parent. Social and economic factors such as poverty and a dysfunctional family may provoke chronic stress, leading to alcohol abuse and illegal drugs. This type of behavior may be passed from generation to generation. However, there seems to be no scientific evidence that there is a genetic modification or alteration passed on from parent to offspring. I have not come across any epigenetic research done on this mental brain disease. There are some social scientists that correlate drinking and gambling to cultural and behavioral tendencies, but a genetic trigger is yet to be found. However, in the brain, there are groups of cells like nucleus accumbens and the ventral tegmental area that reward repetitive behavior in humans. Several scientific reports have supported evidence regarding repetitive behavior in humans provoked by rewarding brain cells. Excess alcohol drinking, gambling, as well as consumption of drugs like heroine lead to addiction. Some lay persons and professionals claim that the brain is hijacked by rewarding brain chemicals. Nevertheless, it should not prevent me from my focus on memories.

"Memory is probably what gives each of us our highly individual and personal identities."[3] We begin to encode and store memories right into our brain's synapses from very early in our life. I believe the first memories we begin to store are those of our mother's, especially if the baby is breastfed. The image or memory we incorporate is stronger not only for the physical contact but also for the hormonal attraction. There is a hormone called oxytocin that promotes strong mother-child bond. The mother- child emotional and biological bonds are unequivocally

3. Nancy C. Andreasen, Brave New Brain, Oxford University Press (2001)p.60.

stronger than any other. The intimacy and closeness of baby and mother experiences along with the mother's character and emotional makeup transforms the mother into a tender and selfless object to safeguard her child's safety and growth.

It is perhaps an unconscious attempt to splice or join the broken umbilical cord that was cut off from the placenta. The baby captures the intensity of multiple emotional vibrations coming from the mother, reinforcing the unique intense mother-child bond. The mother's memory will be permanently stored in the child's brain. As the child grows up physically and emotionally, he/she expects the mother to rescue him/her at any cost.

I must add there may be circumstantial triggers that may interfere in this bond between mother and child. The content and intensity of memories stored in the child's brain will greatly determine his/her mental, physical growth, and feelings of wellbeing. A growing organism in need of nutrition, protection, and social interaction, the baby makes many demands. The caretaker makes all attempts to provide; however, it may not be possible to satisfy the child's demands at will. In most cases, there is an accommodation between the child's demands and the mother's ability and willingness to provide it. Consequently, as brain cells are growing through dendrites and axons, thus forming synapses, I believe the process of pruning of dendrites and axons begin to take place. The child needs to obey the mother and father's behavioral rules. These rules will be permanently stored memories that will guide the child to a successful life. Some early memories will be lost through the pruning process to give way to loving, compassion, and caring memories.

I will name these memories as love and anger stored memories. A child's anger memory need not to be realistic nor based on child neglect. The mother may have attempted to provide all the child's needs, but the child cannot understand the mother has other chores to accomplish. A child may react to parental restrictions and discipline with undue anger, which may be translated into guilt during adolescence. During my practice as a psychotherapist in New York City, I came across several cases of this nature. The little crying child inside the adolescent had not outgrown his infantile needs or feelings of neglect. There were misplaced anger, frustration, and guilt to be worked out and solved during therapy.

Generally, memories are stored at postsynaptic level. The process of storage normally begins in the soma or body of a neuron traveling along the axon. It is an electrical impulse discharging a chemical message at the space between two neurons called the synapse. The chemical message will be received by a receptor molecule at the next neuron. A chemical message could be a neurotransmitter like dopamine, glutamate, or GABA. GABA is an inhibitor while glutamate is an excitatory neurotransmitter.

Chapter IV

Dopamine

Dopamine is a neurotransmitter that makes you feel happy. However, an excess of it may ruin your life. For example, a friend of yours may invite you to have a good time playing machines at a casino. Let's imagine he loaned you $20.00. You played the first ten dollars within the first thirty minutes and lost it. You decided to move to another machine. You pushed a half dollar into the slot, pulled the handle, and got back a hundred dollars in quarters and half-dollar coins. In addition to money, your brain rewarded you with the neurotransmitter dopamine. You had dopamine governing your body. Most probably you began jumping, laughing, and calling your friend to your side. You hugged him and thanked him for this blessing. Brain cells in the nucleus accumbens and the ventral tegmental area, these two feel-good brain areas, were actively releasing its chemicals. You went home with $500.00 in your pocket to celebrate it with your wife. Most likely you could hardly wait to get back to the casino and make more money, and perhaps you did. During your third trip to the casino, you did not do as well as the first two trips. However, you were making money. Although the third trip was not as good as the first one, you still had money left. You were convinced that you would win the big jackpot on your next trip. You did not want t to listen to your wife or your friend encouraging you to stop and enjoy what you already had. No, you wanted more! Endorphins and dopamine had won you over.

What you were not aware of is that you had two groups of neurons already activated and ready to release feel-good chemicals even before you arrived at the casino. This imagined episode may end up as a gambling addiction. I will have to classify this memory as a toxic memory and shelve it with previous memories that may provoke anxiety and anger.

Scientists have developed sophisticated brain scanners like functional MRI that shows you how these brain cells get activated with just a picture of a casino with slots machines. We love to have a culprit to blame when it comes to an addiction. We tend to look around for friends, suspecting family members, fault a stressful job, or attribute a hereditary family trait. However, we tend to overlook the most powerful organ in our body, the brain. There are areas in our brain that respond to pleasure, reward, and decision-making. Brain cells in these areas communicate with each other using chemical signals known to us as neurotransmitters. We experience pain when touching a very hot object, and we quickly respond by withdrawing our hand. Similarly, when we have a pleasurable and exciting experience, the neurotransmitter dopamine is released in significant quantity.

When we repeat a similar pleasurable experience, those cells that released dopamine seem to have stored memory that triggers the release of dopamine again. Before you realize it, you will be craving to repeat such pleasure-provoking experiences. If you were taking heroine, it will hijack your brain despite your efforts to stop. I have had people on their knees begging for a cure for their brain disease, addiction. Heroine is the most addictive drug known. It mimics the brain's own endorphins, the natural opioids that induce pleasure and reduce pain. The brain responds by reducing the sensitivity and number of its opioid receptors, so that more of its drug is needed.[4]

It is hard for me to call it memory. We become victim of brain circuits and its decision if it is a decision. I would like to call it a brain

---

4. BBC, *The Amazing Brain* (January 3, 2017), 66–69.

deficit. I was victimized by a teacher while in secondary school. It became a very toxic memory that haunted me for many years. Now it is called PTSD. It hijacked my brain; I could not get rid of it by myself. The fine line between different types of memory seems to blur as I go looking for my primordial memory.

The formation of memories has intrigued philosophers, theologians, historians, poets, and many lay people in many countries and throughout time. The Greek historian Herodotus made an oral history (memories kept by word of mouth from generation to generation) into written stories. One of which was of a fight between Greeks and people in the city of Troy. While reading this story or history of two great people fighting to defend honor, country, and motherland, I felt I was involved in the fight. You find memories of love, anger, vengeance, deceit, betrayal, loyalty, pain, and more. It involved men, women, and children. According to some archeologists, this battle took place around three thousand years ago, yet we enjoy reading of it. It deals with human emotions and feelings. In the story, you find love of a man for a woman, love of a mother for her child, and love of a father for his son. I love to read it because it relates to my childhood experiences and perceptions of people around me.

With good history memories, I feel I can make better judgment and decisions at present time. I am in a better position to investigate my unconscious and conscious mind and enjoy a more satisfying life. Remembering the battle for Troy as a great memory of love, vengeance, and deceit brings me back to our present. We have been engaged in a war for over a decade in the Middle East after we lost several thousand innocent people in a terrorist attack in New York City. During the terrorist attack on the Twin Towers, I felt angry, sad, and helpless. I was on the roof of my home in Brooklyn, watching the second building collapse, and the ashes from the building covered my head and shoulders. I could not help it, but I became aware that tears were running down on my cheeks. These two tragic memories, Troy and the Twin Towers, make me place time in a linear fashion, a time space of three thousand years. I place Troy in northwest Turkey. I read books and magazines on history, religion, philosophy, civilizations, arts, and of course, wars and science.

While I am reading history books, I come across Persia, Babylon, Athens, and Jerusalem as the cradle of multiple religious beliefs. I wonder how my brain stores those countries and cities. Are there specific shelves in my brain to store big names like Jerusalem, Babylon, Egypt and their pharaohs? Are Achilles, Zeus, Socrates, Plato, Moses, and Jesus stored differently or share the same box in a cluster of neurons? Are time, places, and people stored separately? I use those people, places, historical events, and geographical locations to place and orient myself mentally and physically. I need to know where I am in time and place in relation to mankind. To do that, I must rely on my memory bank somewhere in billions of neurons and their synapses. I must correct myself and say, stored in molecules in my brain's cells.

Besides memories related to very important people and great cities, there are emotionally loaded memories stored in specific regions of my brain. We have a region in our brain that is the seat of emotions; it is called the limbic system. It was formed in our brain a long time ago before our neocortex had evolved. Can the neocortex be the multiple groups of brain cells chosen to store emotions of love, anger, fear, jealousy, enviousness, and hate? Most people in the medical field, especially neurologists, know that the hippocampus is the first location or stop for memory storage.

The following is a sad story. We learned about it while a brain surgeon was trying to help a patient H. M. who had severe epileptic seizures. H M. had seizures every day that prevented him from enjoying a normal life. This brain disorder did not respond to medication or any other therapeutic technology available at the time. There is one hippocampus in each temporal lobe. The medical team in charge of H. M. recommended removal of both hippocampi after the removal of the first one had failed to stop the seizures. After the second hippocampus was removed, H. M.'s seizures stopped. However, as sad it is, H. M.'s ability to store memories was stopped after the second hippocampus was removed. He could remember and talk about episodes in his life before surgery but nothing after. You could take H. M. to watch and enjoy a baseball game, and after the game was over, he had no recollection of it. You could take H. M. out for hamburger dinner, and on the way home, if you asked him if he enjoyed the dinner, H. M. had no recollection of ever going to the hamburger place.

People who have suffered severe brain injuries in car accidents and combat soldiers have had similar problems. A very sad experience established forever the significant importance of the hippocampi.

Chapter V

Hippocampi

The hippocampus is generally considered by brain scientists and the medical professionals as the first stop for immediate or everyday working memory for further distribution in the brain. We must stop and seriously think about the function or functions of the hippocampus in memory storage. If H. M. could not remember having watched the game just five minutes after the game was over, how would H. M. emotionally respond if a car ran over a small child on the street? Did he become immune to pain, anxiety, or any other environmental threat to his life and others? Can a drug numb both hippocampi of combat soldiers to be fearless, ruthless killers? We must bring these remote assumptions back home to our memory storage. How does an emotionally charged emotion travel from the hippocampi to the synapses for storage? Where in the brain, besides both hippocampi, are those synapses located, and how can we find it?

Structure determines function. Is it possible to use memories' molecular structures to follow their pathway home or to a storage room? There have been several research attempts to trace the structural changes in the nervous system produced and left behind by long-term memories. Researchers can use dyes and isotopes to follow tracks of molecule signals from cell to cell in lower organism. However, the human brain is not a toy to play with. I imagine that long-term memories with a strong emotional load attached to it would be easier to track than any other memory. I assume that the starting point to track any memory must be the hippocampi. I already suggested that tracking an emotionally loaded memory must implicate the limbic system that makes connection with the prefrontal cortex.

Neurosurgeons have reported that while they were attempting to introduce electrodes in the striatum (putamen and caudate) in a Parkinson's disease patient, the wire used to introduce the electrode unintentionally came in touch with nearby neurons. Consequently, the patient began to cry aloud with suicidal ideations. Once the wire was moved to its anticipated location, the tears and crying stopped. It shows us how sensitive neurons are even by the slightest touch of an external stimulus.

Parkinson's disease is a neurodegenerative brain condition provoked by loss of dopaminergic neurons in the substantia nigra. It manifests itself clinically in the form of characteristic motor defects like tremors. At ease, the patient's hands shake constantly, causing difficulty getting up from a sitting position or taking a cup or spoon to their mouth. Their emotions are not easily visible, which provokes people to call Parkinson victims poker face. Most symptoms of Parkinson's disease come around age sixty, but you find cases of an earlier onset around fifty years old. There is strong indication for a hereditary past history in the family. Gene therapy and stem cell therapy have been considered as possible treatment for Parkinson's disease. Electrode implantation deep in striatum has been used for several years now. At one time, induced pluripotent embryonic stem cells were seriously considered the most viable treatment for Parkinson's disease.

I have named several loci to find long-term memories storage; the next step is how these memories are formed. At present time, there is a consensus among researchers that memories are processed at synapses. My next questions are the following: How are memories formed? What parts of each neuron are implicated in the process? How many neurons are involved in memory making? How are the DNA and RNA macromolecules engaged in this complicated process? What parts or vesicles inside the cell provide the energy to form a new molecular structure for an episodic memory?

Researchers from around the world have implicated the right hemisphere of the brain in the storage of long-term memories related with time and place or location. Also, there is a worldwide consensus among researchers that the brain's left hemisphere is dominant for speech. There are two areas in the human brain that are specifically related to the spoken word. One is named Broca's; the victim cannot speak. The other is named Wernicke's; the victim can hardly be understood.

The next step would be to take memory to the laboratory. The best object of study would be another human being, but we cannot do that with humans. We could do it with lesser humanoids like chimpanzees or monkeys, but it is illegal in the Western world. Perhaps we can do this type of research with birds, flies, worms, snakes, or with alligators. We have used a mouse as our best toy of research for multiple purposes. There are multiple variants to consider before grabbing the next research object. How can we isolate a single neuron or multiple neurons for an extremely complicated job we intend to do? Where and how can we handle the object of our research? What type of machines and technologies do we need to employ to do a job that ultimately needs to be replicated to be verifiable? It seems obvious to me that it is work best done by a team.

We have named several classifications of memory that must be studied. We have working memory that we use for daily communication. In addition, there are short-term memories that are readily available when needed in daily chores. The separation of these two memories seems arbitrary and often confusing, but there is a difference in time and place. Next, we come across long-term memories that may require temporary structural change of neurons. And last, I will place toxic memories that provoke permanent anatomical change in the implicated memory. Eric R. Kandel, Nobel Prize winner for his work on memories wrote, "Short-term memories produce a change in function of the synapse . . . Long-term memory requires anatomical changes. Repeated sensitization training (practice) causes neurons to grow terminals."[5]

---

5. Eric R. Kandel, *In Search of Memory* (New York: W.W. Norton, 2006), 215.

In these three last sentences, I find a treasure chest of wisdom and brilliantly researched discoveries. It elucidates and clarifies for us the processes of memory formation and neuronal alterations or changes. It strikes me most that in long-term memory, the neuron is structurally altered. The neuron grows new terminals. We must assume that a memory capable of forcing a neuron to undergo self-changes must provoke behavior changes too. This anatomically changed neuron can have the potential to attract more signals than its neighbors, thus altering normal behavior. The hypersensitization of this structurally modified neuron may cause it to easily respond to external stimuli and form an abnormal response and behavior.

Also, if you change a person's behavior, you change his personality. This is the main reason that justifies my argument that our criminal system should invest its money in providing successful behavior-modification therapeutic technology in prison wards. Recidivism is horrible in our present justice system; it is a revolving door for a numerous sections of our prison population. Perhaps if proven therapeutic, modalities reinforced by learning new behavioral skills, the number of our prison population coming back to prison for shelter, food, and socialization would be vastly reduced.

Our educational system could also benefit from working with scientifically proven models of successful technologies. These technologies must address not only our social medium but also our genetic and epigenetic heredity. Survival tools from most urban ghetto areas do not equip our youngsters to enjoy average American lifestyle. I must make a strong statement based on my experience working with both, recidivists and school dropouts. Their brain neuronal networks and memory folders and archive files seem to be full of tools for their neighborhood but poorly adapted to the main American competitive lifestyle. We can train them to fight terrorist wars, but coming back from those areas poses a bigger challenge than before they went to war. We already have this problem at home. Europe, especially, France, England, Belgium, and Spain, has been a victim of ill-trained ghetto returnees. Russia's experience in the Chechen Republic seems to be another case to study.

All that I have discussed here has been centered around memory and the hippocampi. Scientists, meaning neuroscientists, ever since Cajal's neuron doctrine in the early twentieth century have known that the hippocampi can grow new synapses. Can we modify these new synapses to reject harmful and risk-prone incoming memory signals? Stop. I anticipated your objection. The censorship must have community involvement to be useful and approved by a large majority of people benefiting from this learning experience. Their children will be better prepared to enter higher institutions of learning and acquire the best employment skills to succeed in our society. I understand how difficult it may sound to a section of our society, but unemployment and violence in general are calling for changes we may not be prepared for.

During our economic depression of late 1920s and early 1930s, crime and violence were driving our country into a catastrophic conclusion. President F. Roosevelt introduced legislation that was approved by our Congress to get federal agencies involved in public works providing jobs for millions of unemployed Americans. Ironically, he was strongly criticized by a good section of the American society. I recall reading some newspaper articles calling our president a communist-inclined person. However, despite some awful criticism, President Roosevelt won reelection in all but one state of the union. He did not force anyone to vote for him, but the wisdom of common sense prevailed.

Their brains may be compromised when it must make intelligent and well-reasoned decisions. Neuronal signaling and connections would be altered, thus preventing the proper and normal convergence of messages reaching the decision-making group of brain cells. I think of toxic memories when I come across a person suffering PTSD. In this case, I will be talking with an individual whose brain cells have undergone anatomical changes. That person's behavior could become unpredictable. I must be ready to respond to a behavior not normally expected or typical of ordinary people. You may have family members and friends who have served in the American armed forces in Vietnam or in the Middle East or other areas of conflict. Some of those soldiers came home with serious physical and mental problems. The Veteran Administration has a system of hospitals throughout the country to care for them. However, in many instances, mental problems are not adequately diagnosed at the time of discharge or separation from the armed forces, and our veterans do not receive the kind of care and treatment they deserve. Consequently, they may resort to abusing illegal drugs for self-medication to alleviate their mental anguish. At a personal level, my oldest brother, a WW II veteran, did not receive any treatment in part because of ignorance but basically because the VA did not follow up upon his discharge. He had been wounded in a leg and received sedatives for combat fatigue. In the late 1970s, I worked at a veteran hospital in a large Northeastern city of our country. There was not much interest in following up with veterans discharged from the Vietnam War who complained about problems in their head. I did care, and I received a commendation from the hospital.

My point is that in general, we do not pay as much attention to brain diseases and mental disorders as physical problems. Traumatic mental disorders as well as brain diseases can have multiple causative factors that go unnoticed or ignored by the general population. It is my impression that the tendency to pay less attention to brain diseases and mental disorders is partially based on poor diagnostic tools. We can easily x-ray a fractured arm or leg and get a clear picture of the problem. Medical doctors have at their disposal sophisticated scanners, blood testing tools, as well as many other diagnostic tools that help them make appropriate diagnoses. However, tools to locate anxiety, pain, intrusive thoughts, suicidal and homicidal ideations, hallucinations or delusions in the brain of a person at risk are not available at present time. The most recent discoveries of anatomical changes in a hyperactive neuron leading to toxic memories like PTSD need the attention of all mental health professionals as well as the public.

I am looking at a picture in one of my selected books of psychology and psychiatry. I cannot help but deeply admire neuroscientists Santiago Ramon y Cajal, Emil Kraeplin, Wilhelm Wundt, Sigmund Freud, Alois Alzheimer, Ivan Pavlov, and Brodmann. Cajal saw the brain composed of separate individual cells in communication with each other. He must have had extraordinary ability to observe and draw brain cells, including dendrites and axons. E. Kraeplin was a practicing neuroscientist even before the neuron was recognized by the medical profession. Dr. A. Alzheimer performed an autopsy on one of his female patients, trying to find out why she had lost all her memories. To his surprise, he found that protein deposits known to us as amyloid plaques and tau tangles had accumulated in regions of her brain controlling memory and thinking. Amyloid plaques anchored between neurons' synapses block their communication while tau tangles make the inside of each neuron a huge mass of twisted strands of vesicles. I have seen this brain disease in people begin as early as forty-seven years old. As we grow older, the chances of being inflicted with Alzheimer's almost double. This horrible disease has been in the mind of Yoshinori Oshumi, Nobel Prize winner in medicine for 2016. Talking with a reporter from *The Washington Post* (10/03/2016), he said, "I found that we have an ongoing renewal process without which living organism cannot survive." He continued, "It is a key mechanism in our body's defense system that involves degrading and recycling parts of cells, known as autophagy. This process plays an important role in cancer, Alzheimer's, type 2 diabetes, birth defects, and numerous other devastating diseases." Inside a cell you see a vesicle known as lissome. One of its functions is to go out and rob debris from intercellular medium and bring it inside the cell for processing and distribution.

I should note here that there seems to be some confusion between a diagnosis of senile dementia and Alzheimer's. Before Alzheimer's disease was discovered, senile dementia was the preferred professional diagnosis. However, the end of the nineteenth and beginning of twentieth century saw the brain come out of the realm of witches, evil possession, and heavenly punishment and into the scientific world as the organ of the human body controlling our behavior. This marked the path for future brain researchers to place psychology and psychiatry out of religious and philosophical innuendos. The study, diagnosis, and treatment of brain diseases became a distinctive discipline of the medical profession. We are lucky their books were not burned in public squares or the authors sent to prison.

Following the discovery of anatomical changes in neurons by Eric Kandel, he added, "The number of synapses in the brain is not fixed—it changes with learning." To understand Kandel's quotation, I must go back to the classification of memory into short-term memory and long-term memory. Short-term memory is used for regular-normal communication. These are memories retrieved by the hippocampus with great ease. Most, if not all, are stored as a molecule at synaptic level. The more frequently we make use of that molecular memory, the stronger the potential for contentlike associations triggered by similar stimuli. However, the potential to widen and strengthen the synapse of a short-term memory depends, in part, on what we mean by *learning*. Does the rote of daily living done without any thought of meaning enrich or provide strength to its synapse? Similarly, is it fruitful to memorize a group of numbers, a poem, or the dates and names of great men in history? As the author named above indicates, long-term memory requires anatomical changes. It seems to me that learning requires consistent activation of multiple neurons making positive connections with groups of brain cells to accomplish a shared goal or need. The need to move my arms or legs is first made in the brain and then executed by the rest of my body. It is written in our genetic code but needs cellular growth, specialization, maturation, and execution. A child's learning would be extremely limited if exposed only to rote memorization.

The sooner a child's brain is introduced to rich environmental stimuli, the greater the potential to attract and activate several brain cell network in a common goal. We may assume that it would also help to avoid getting stuck in a repetitious hyperactive synapse leading to a toxic memory like PTSD. I consider a posttraumatic stress disorder a violently formed unfinished memory, processed when an individual is overwhelmed by an assault and does not have access to the necessary tools for self-defense. It is not my intention to be confusing with multiple complex memories. I cannot ignore unconscious memory formation and storage.

I have read several authors claiming that the volume of our stored memories is immense. They compared it with a huge iceberg. The conscious memory is just the tip of the iceberg while unconscious one is underwater, not seen by people. Basically, there are two processes involved in long-term memory storage: suppression and repression. Repression, according to Freudian theory, takes place at an unconscious level. Repressed memories generally produce anxiety, fear, indecision, insomnia, headaches, and several other mental and physical disorders. In most cases, these problems respond well to psychotherapy. My concern with repressed memories is that it may have formed anatomical changes in a brain neuron. My experience in psychotherapy with patients who have had what I have called toxic memories is very complicated. First, the patient needs all support I could provide in the present, here and now, environment. There was room for me to go into a long personal and family history. However, I had to take each individual back in time to the formation of the trauma. I had to gain the patient's confidence to trust my therapeutic skills challenging the object or objects triggering the traumatic episode. I had to convince the patient that I was ready to join him challenging the offensive object. I was there for him when needed.

However, before getting ready for the battle, we went over naming possible available tools at his or her disposition not used at the time of the assault. I believed it was necessary for him or her to deactivate the overreacting cell or group of cells involved in the trauma. There was a lot of anger, fear, and feelings of helplessness involved in each individual case. There is no need to emphasize that each individual person must be fully equipped to face the threatening object before he or she enters action. After a long psychotherapy practice, I came across the inevitable truth that I was dealing with anatomical changes in a neuron synapse. I came across E. Kandel's discoveries: "Short-term memory produces a change in the function of the synapse . . . Long-term memory requires anatomical changes . . . causes neurons to grow new terminals."[6]

While in training, it was no wonder I was hitting my head against a hard rock with limited success until I found E. Kandel. No verbal therapy, meaning talking about a trauma without confronting the aggressor, was helping me get rid of my tormenting, torturing, anxiety-provoking memory. It seemed to me that my situation was getting worse

6. Ibid.

as I came out of my therapist's office. I did not receive any relief from pain, anger, fear, and guilt. On the contrary, I was beginning to blame myself. I felt I was not intelligent and strong enough to understand the therapist's expectation of me. I felt I was too slow to understand the dynamics involved in traditional psychotherapy. However, the perplexing difficulty, the crux of the matter was that I was engaging the wrong part of my brain to knock down a wall in my brain. I was using my tongue, my mouth, which are connected to my brain's frontal cortex, while leaving unengaged the emotions and motor cortex that are vital in any trauma.

The amygdala, hippocampi on each hemisphere of my brain as well as the limbic system and motor cortex were, in a way, suppressed for the sake of talking. Signals from my eyes, ears, and skin leading to and connecting the amygdalae, hippocampi, limbic system, and sympathetic nervous system—all frontline groups of brain cells instantly engaged in any trauma—were placed by the side during a verbal therapeutic session.

As E. Kandel found out, new terminals in the neurons had grown with permanent long-term memory leading to a toxic memory. I did not know at the time that I needed a stronger and more effective tool than my tongue to bring down a wall built by new synaptic buttons. PTSD victims need to be active participants while challenging their aggressor. They must be well equipped with appropriate tools to be used at the proper time and place. If effective self-defense tools were lacking at the time of the trauma, the victim must be willing to use modern and imagined models of defense for self-survival. This is an encounter with a fearsome and dangerous object that needs to be overcome or destroyed. It is an unfinished assault memory that needs to have a complete and self-accomplished closing and a permanent end. A frightening and overpowering traumatic memory tends to appear spontaneously and repetitiously, provoking fear, anger, feelings of powerlessness, and even suicidal and homicidal ideations. Thus, the victim is in dire need of support to be able to get ready for the eventual confrontation with the aggressor.

The victim needs practice mastering the necessary tools now at his disposal. For example, a young lady may carry in her brassiere a blinding laser or spray for self-defense. She used a dummy one for practice at my office. There are more sophisticated tools for self-defense a lady can carry without raising any suspiciousness. We must have in mind that PTSD or toxic memory must have formed anatomical changes in neurons; therefore, the victim must be well trained and strong enough to alter, modify, or destroy the violent and aggressive object that has hijacked her brain. Otherwise, the toxic memory will stay in her brain for life.

Chapter VI

Philosophers

Philosophers, educators, physicians, and recently, neuroscientists have joined the team interested in elucidating our understanding of our brain and memory formation. There are two clinical cases of memory deficit that have always intrigued me: H. M.'s hippocampi removal and Alzheimer's disease. H. M. kept his memories before having both hippocampi removed, but he could not store and retrieve any memory after surgery. As far as I know, H. M.'s neurons were alive and responding normally if it was not necessary to change one topic of conversation for another. Simply, there was no massive destruction of neurons except in the hippocampi.

In Alzheimer's disease, neurons are blocked by protein blockages, and the inside vesicles of each neuron become entangled, preventing all the physiology that normally takes place in a normally healthy brain cell. My experience with Alzheimer's patients during the advanced stage taught me that each patient needs total care by a person who cares and understand this brain disease. The victim loses all her/his memories. The victim urinates and defecates in bed or wherever she/he may be at the time. The victim needs to be fed and dressed, for she/he does not know the need to wear clothing. During winter, when we need to dress appropriately, an Alzheimer's victim may walk out of her home naked or inappropriately dressed and freeze to death. This type of saddening and painful situation has called the attention of some lay people and professionals to say that we are human beings because we can store memories. However, it does not seem to me a complete and satisfactory characterization of memory. Despite an Alzheimer's victim's loss of her/his conscious memories, the organism continues to be alive.

I shall continue my search for and characterization of memories. I have named some traumatic memories that incapacitates the victim. I am in an impasse, trying to decide on toxic memory formation or repressed memory formation. Both memories are very painful memories that interfere with our daily life and health. These memories may prevent us from enjoying a happy relationship with our partner and family members. Our academic achievement is compromised by these persistent and recurrent episodic toxic memories that sprout spontaneously. Occasionally, these traumatic memories can be triggered by unrelated signals wrongly interpreted by the victim. The toxic-memory victim may respond with aggression secondary to misguided signals and interpretation. On both cases, at least minimally, the eyes, ears, hippocampi, and limbic system are involved. However, I am aware that I am a long distance from the road that may lead me to the loci or original place that forms a human memory. I hesitate to go on my research on memory because there are so many variables and connecting themes that all seem to converge in the process of memory processing. They appear to be infinite, like stars in the universe. However, this exaggeration is unnecessary. It is a difficult theme to handle, but there are sophisticated and efficient technical tools to achieve my goal.

Among many things that stop my search are interesting thoughts like conscious memory. *Conscious memory* is also known as declarative or explicit memory, which comes almost spontaneously. Also, there is implicit memory that is not intended to surface during a friendly conversation but comes out provoking embarrassment. Can I classify these types of memory a toxic long-term memory or suppressed or repressed memory? I made those connections because as a child, I wished I could make my parents disappear when they did not allow me do things my way. I have a lot of childhood memories that were full of wishful thinking. I hope that during the pruning process of dendrites and spines on my neurons, a good job is done cleaning it up.

For a memory to become long-term, it needs to be repeated several times. In the end, it may become toxic if painful to the victim. Recently, newspapers and television have reported several cases of bullying in school, which provoked the victim to commit suicide.

Whether I like or not, it seems that I must go back and look at my brain neurons for the initial process of memory-processing cycle. It is the neuron, its dendrites, and axons that are responsible for the formation of our memories. Each neuron may have several thousand dendrites around the cell body but only one axon. Dendrites may look like fingers around the body of.the neuron. All receive messages from other neurons. They are chemical messages carrying a positive or negative value. Coming out of the body of the neuron, there is a long myelinated tube or pipelike structure that is propelled by an electrical force that will deliver a chemical message at a synapse. A synapse is the connecting space between two neurons. The chemical message at a synapse will be received by a friendly molecule at the recipient neuron. Each neuron body will receive thousands of these chemical messages. The sum of positive and negative messages to the neuron body will determine whether the cell will fire or not.

The location and distance of each synapse from cell body is another variable that influences the neuron's body physiology. When two neurons communicate with one another, it is a chemical exchange propelled by electricity. So we are dealing with positive and negative molecules, atoms, and ions. We become involved in the intricacy of the development of human life itself.

The two basic acids of all living organism, DNA and RNA, with all the necessary elements, form part of the journey that I innocently undertook. Without doubt, despite multiple functions of all our body parts, there is the neuron at the center of memory processing. The neuron, like every cell in my body, is surrounded by and washed by extracellular fluid that is loaded with an abundance of ions. Ions are electrically charged atoms such as potassium, chloride, and sodium. Sodium ions are positively charged while chloride ions are negatively charged. (Our table salt is a compound of sodium and chloride.) These positive and negative ions play very important functions during the axon physiology, forming and delivering a chemical message at a synapse. Further exploring the physiology of neurons, the inside fluid or cytoplasm of the cell is very rich in proteins, which are negatively charged. This establishes a working balance with potassium ions, which are positively charged.

As I mentioned above, an axon is a hollow structure with tiny special openings called ion channels. There is a dynamic in and out flow of potassium in the axon. As potassium moves out, it leaves an excess of negative-charged proteins. Here we can see with better clarity the attraction of positive potassium playing its determined role coming inside the axon attracted by the negative-charged proteins. During this dynamic in-and-out process, I must include the movement of liquid from a higher concentration state to a lower concentration state. On a flat surface, I connect by the bottom two containers with a capacity to take in twenty-four ounces of water each. I fill up container number one first and then open the faucet, allowing water to flow freely to container number two. There is a gradient force from water going to container number two from number one. There are other factors involved in this process that I will not discuss here.

I have mentioned potassium, sodium, and chloride involved in the delivery of a chemical message to a synapse. These are electrical and chemical processes with the potential of becoming engaged in the process of forming a memory. However, I cannot call them memories of any kind. They are just reactions of different elements necessary for my organism to function normally. Even the process of protein formation by genes is just a chemical and electrical reaction. My skin reacts to heat, cold, and touch; is there a memory formation or just a reflex reaction? Similarly, is my organism's innate defense system a spontaneous organic reaction, an epigenetic inherited memory or an unconsciously stored old memory? All current memory research and past clinical practice and accidents point to the hippocampi for memory storage and retrieval. The amygdala, on both temporal lobes, also store emotional memory. It is a survival brain cluster of specialized cells. I can also refer to unicellular organisms without a brain system whose behavior is just a normal reaction to an environmental stimulus.

Occasionally, I go to my backyard at home to look for and play with insects and earthworms. Using my globes, I take a few worms and place them on a flat surface and watch how they respond to different stimuli. I may use heat or cold as a stimulus. Also, I have used sugar, salt, vinegar, lemon juice, garlic, and combination of them to determine behavioral responses as well as survival skills.

Chapter VII

E. Kandel on Santiago R. Cajal

I am going to leave dormant, for the moment, my curiosity on electrochemical reactions on brain cells, particularly, in the cell's axon. While looking for memory formation, I stopped at one of E. Kandel's monumental research on the subject. He wrote, "I found the same specificity of connections between individual neurons that Santiago R. Cajal had found between populations of neurons. What's more, just as neurons and their synaptic connections are exact and invariant, so, too, the function of those connections is invariant."[7] Cajal had proposed that the connections between neurons known to us as synapses are altered in some fashion when memories are formed. I understand that it means that when we learn anything, we form a memory that will change the structure of the synapse, creating a niche of synapse hyperactivity. This niche of hyperactivity may attract positive or negative behavior forming good or bad habits. At first, the same specificity of connections between individual neurons opened my eyes as never before. It means a pleasant, enjoyable, and love-provoking connection may influence me to look for and wish to repeat that specific neural connection.

For example, when I first tasted a very sweet and delicious pineapple my father had planted on his farm, I could not help it but sneak out of his sight to eat his treasured pineapples. I knew he was on guard to

7. Eric Kandel, *In Search of Memory* (New York: W.W. Norton, 2006), 194.

catch the neighbor who was destroying his selected specimen for future plantation. When I was caught, I got what I deserved: a good whipping. The taste buds in my mouth provoked a behavior that was becoming an unstoppable habit, and habits can become an addiction. I must introduce here dopamine, the feel-good neurochemical that is always ready to make us repeat, among others, an enjoyable mental thought, a delicious pineapple, or a kiss from our sweetheart.

Now, I am in memory formation again. I can talk about good and bad memories. The first bad memory to come to my mind is "Thou shall not steal" perhaps because of the whipping I got from my father. I also keep good memories of my father. The strongest memory of my father began as a short-term memory. Each evening, he would milk a cow to give me healthy fresh milk. After several weeks and months of drinking milk at the same time from the same cow, it must have become a long-term memory for me to be able to remember it over eighty-five years later. I am going to speculate that every evening when I saw my father coming in with the same cow, my mouth began to salivate in anticipation of a delicious cup of milk. I keep a good picture—memory—of the cow and my father in my brain.

My friend Peter whispered a sour comment into my right ear. He said, "It sounds like Pavlov's work with his salivating dogs."

How do I recall that sweet memory of my father feeding me milk after so many years? Generally, there is the ability in my brain to encode, store, and recall. The first place to store memory is the hippocampus. I already mentioned to you the case of H. M. His hippocampi on both sides of the brain were removed, and he could not remember anything from after surgery. When the hippocampi are removed, the person loses its ability to store and retrieve memory. However, mechanical and automatic memory like pedaling a bicycle is not forgotten. We cannot remember anything if we do not code, store, and recall.

Where does the memory go after it is received by the hippocampus? Experiments done with PET and MRI scanners show high neuronal activity in different regions of the brain. For example, if the memory is emotionally loaded, the limbic system will show high neuronal action; if the memory is visual, the occipital lobe will be active.

We already know that the left hemisphere of our brain is dominant for the spoken word. The idea of functional localization in the human cortex was presented at a meeting of the Anthropology Society in Paris in 1861 by Pierre Broca."[8] There are two classical areas in the left hemisphere that are described and known as Broca and Wernicke areas. If the Broca area is destroyed by accident or illness, the person's ability to speak recognizable words is gone. On the other hand, if the Wernicke area is destroyed, the person's ability to speak in a manner that we can understand is gone. The victim speaks, but their words do not seem to connect with each other. Despite this observation on memory storage after making a stop in the hippocampi, we have to say that the entire brain is responsible for memory storage.

Moreover, another level of memory storage is needed when there are multiple forms of memory in terms of time, space, emotions, financial need, driving an eighteen-wheel trailer truck, or a sniper in super-alert mental state. All these memories need to be sorted out, classified, and sent to specific folders in an archive. Each one these steps must have its own memory to achieve its goal.

For the benefit of clarification, let's discuss a sunflower plant. Where would my brain store the noun *sunflower* after making its first stop in the hippocampi? Would it go in a folder reserved for colors, or will it go in a folder for shapes like the Sun? "No, no," said, John, "it probably goes with food because of its seeds. I eat sunflower seeds almost every day and, it makes me romantic."

Jane, who was reading a book on behavior, interrupted us, saying, "I protest you giving sunflower so much attention and power. I hate sunflowers. Mike, my boyfriend during my second year in college, was a sunflower-eating maniac who tried to force me to live on that damn thing."

You, dear reader, may have a folder to store *sunflower* after this unexpected experience with John and Jane. There are many variables to consider before we decide to speculate about memory storage. However, as I mentioned above, there are regions in our brain that are

---

8. Semir Zekir, *A Vision of the Brain*, (University College London), 17.

function specific. Jane's brain had stored *sunflower* together with the emotion, anger. She verbalized strong dislike for sunflower along with her boyfriend. She used strong and emotionally loaded accusations using words I dared not to print here. Having said that, it seems to me that there is no objection but conclude that Jane's brain must have sent the *sunflower* memory to the limbic system. *Where* in the limbic system is another issue worthy of exploration. I have always associated that region of my brain with anger, rage, violence, criminals, abusers, ugly bloodthirsty monsters looking for innocent victims. It sounds like an inferno. Indeed, hormones from the limbic system make us do things that we later regret. Paradoxically, from the limbic system come hormones that make us brave enough to challenge the enemy even when all circumstances are against our own survival.

Not to complicate even more memory storage and the limbic system, where would you place the following memory? A spiritual father who attended religious service on regular basis came home one evening and confronted a burglar armed with a big and sharp knife threatening to kill his fifteen-year-old daughter. He challenged the burglar in a hand-to-hand struggle to subdue and take the knife away from him. During the struggle, both men went down to the floor with the burglar's knife deep on his chest; he was dead. Where would this spiritual and loving father's brain store this horrible episodic memory?

This memory example that I have described here may sound fictional to you and wish it does not happen in a real-life situation. My experience with people having horrible memories only makes me stronger and more able to deal with it. My past experience as a parole officer taught me a lot. To deal with victims of the Holocaust during WWII and veterans of the Vietnam, Afghanistan and Iraq wars is a big battle.

Memory testing and memory research on insects and other small organism have been going on for a relatively long time. It is the norm of every research to begin with the simple and less complex sample of species. Flies, worms, fish, mice, birds are the most frequently used tools for research. Thomas Hunt Morgan began working at Columbia University in New York City in 1904. He and his students used a variety of animals in their study of genetics; they found the fruit fly

Drosophila melanogaster the most useful.[9] Of course, you may intervene and ask, why not humans? There are multiple reasons why researchers do not begin with humans. First, it is inhuman to use a human being for research purposes. Secondly, it is immoral, illegal, and many other reasons you may come up with. Humans would be too complex and costly specimens to be good tools to experiment with. I may hear an objection from the floor asking, what is the use of a scientific discovery on a fish to human life? I say to you that all living things developed from a matrix living cell. That cell divided and multiplied to form a complex living organism like you and me.

Researchers use the mouse as research tool because it has about the same organs as humans. Doctor Kandel used a fish, the *Aplysia*, to conduct his experiment on memory because, among other things, the number of neurons was small. It was relatively easy to manipulate, observe, and record the *Aplysia*'s behavioral responses to stimulus used by his team of researchers. I need not to mention Doctor E. Kandel's extensive experience working with other tools. "As some investigators studied synaptic plasticity in mammals' brains, others were doing work in simpler systems where one might reasonably hope to make direct connections between molecular and behavioral memory."[10] The research tool indirectly referred to here is the fly *Drosophila*. The researcher quoted above tells us that this fly is useful in the research of memory molecules because is relatively easy to screen for genes that affect fly behavior.

Two distinguished scientists on behavior, Ann M. Graybiel from MIT and Kyle S. Smith from Dartmouth College, shared their definition of behavior and habit while working on memory. "New behavior explored: The prefrontal cortex communicates with the striatum, and the striatum communicates with the midbrain where dopamine aids learning and assigns values to goals. These circuits form positive feedback loops, which help us figure out what does and

9. Edwards Willet, *Genetics Demystified*, (New York: Mc Graw-Hill, 2006), 34.

10. Emily P. Huang and C. F. Stevens, *Molecular Biology of the Brain*, ed. S. J. Higgins, (New Jersey: Princeton University Press, 1998), 172.

does not work in a behavior . . . Habit forms: As we repeat a behavior, a feedback loop between the sensorimotor cortex and the striatum becomes strongly engaged, which help us stamp routine into a single unit, or chunk of brain activity. The chunk partly resides in the striatum and relies on dopamine input from the midbrain."[11]

These researchers concentrated their attention on behavior and habit formation. Basically, they focused their mind on specific clusters of brain cells in the limbic system. The striatum is strongly involved in the Parkinson's disease we already mentioned. This group of cells—the striatum—maintains connection with the midbrain, a very primitive part of our brain in terms of evolution. Once more, we come across dopamine, which becomes implicated in learning. This feel-good chemical, dopamine, in our pre-cortex brain, helps us stamp routine into a single unit, which can easily become a habit. Dopamine would make us feel very happy although the result may be very toxic for our organism's general health. The sensitization of a synapse or groups of synapses would easily force a person under stress to repeat the dopamine-release experience, thus ending in an addiction.

Making use of scanners, researchers can see how habit-forming drugs activates a link in the pathway known as nucleus accumbens. This nucleus is central to addiction. Do not forget that when a person becomes addicted, a change in synaptic structure has taken place. Usually, a memory provoked by a structural change at a synapse will be stored permanently in the brain. This toxic memory would have the appearance of a new sprout at a synapse. It is a new learning memory that is assisted by the very powerful neurotransmitter, dopamine. Dopamine is implicated in the horrible brain disease schizophrenia and several other mental disorders.

I want to alert you that the abovementioned brain loops located deep in our primitive brain are also connected with our decision-making prefrontal cortex. Once these loops are established, dopamine release alone is not the only culprit of addiction. You will have the whole brain,

---

11. *Scientific American* (June 2014): 40–43.

including the prefrontal cortex, dictating your behavior because it also has been hijacked.

For example, a gambler will let his children go to bed hungry while he wastes his paycheck in a nearby casino. Adding to this horrible scenario in mental disorders and brain diseases, our brain's prefrontal cortex is the last part of our brain to gain maturity. Hence, adolescents are exposed to myriad external and internal health risks.

My trip into memory and memory storage has led me to areas where long-term memory, working memory, explicit, and implicit memory overlap with memories based on chemical attraction, physiological reactions, and motor reaction as in pedaling a bicycle. It is not that I want to make my job more difficult than it is, but we were behaving and forming habits long before we learned to speak. Most of our body developed behavior and memory storage on brain loops before the neocortex was developed. It was our brain's medulla, limbic system, cerebellum, and other deep-brain clusters of cells that dictated our behavior.

Chapter VIII

Definition of *Memory*

Memory from our collective unconscious, as psychologist Carl Jung called it, determines part of our behavior. If it were possible for us to repair our memory storage process, where would we begin? Someone may suggest three and a half million years ago with Lucy in Ethiopia. You may suggest seventy-five thousand years ago with the coming of *Homo sapiens*. There are several brain diseases in need of urgent repair. Timidly, we are already doing it on Parkinson's disease patients. We are using electrodes. There have been several reports that Chinese neuroscientists are also advancing this technology. These are new technologies that have been developed during the last half century. I was born decades before J. D. Watson and F. Crick figured out the DNA structure, the double helix. In retrospect, I must have been ten years old when chemist Erwin Chargaff showed us that in any DNA molecule, the amount of cytosine equals the amount of guanine, and the amount of adenine equals the amount of thymine.[12]

I had no memory recollection of Watson's, Crick's, or Chargaff's discoveries until I entered college. I joined a group of friends celebrating Watson and Crick's double helix presentation in New York City as if it were my celebration. My point is that memory does not seem to be restricted to time and place. If my mind does not betray me, Carl

12. Edward Willet, *Genetics Demystified* (New York: McGraw Hill).

Jung's theory of collective unconscious interprets my fear of snakes, particularly cobra and coral snakes, as stored memories when my ancestors used to live in caves or poorly protected huts. In the same vein, small children are afraid of darkness and being left alone. My own experience celebrating the double helix and C. Jung's theory of the collective unconscious leads me to believe that memories may have a genetic component. In other words, I may carry memories that are very old. My brain at the time of birth was not a blank slate. Besides basic survival reflexes common to all living organism, the human brain seems to be equipped to overcome its own physical limitations.

Next, I will present some interesting but confusing definitions of *memory*. "Memory is the faculty by which the mind stores and remembers information." It is called a faculty of the mind. I found several definitions for *faculty*; therefore, I decided to leave it alone. However, the definition calls memory a faculty of the mind, the mind. So secondly, what is the mind? It says the mind stores information. Most neuroscientists believe that the power to store information relies upon the brain cells known as neurons with its dendrites and axons. It is the brain cells with the two acids, DNA and RNA, that possess the code for genes capable of processing protein. And memories are stored on proteins facilitated by electrical and chemical impulses. It is at ribosomes where you can pick up proteins for memory storage. This process of protein formation is possible through chemical and electrical actions and reactions. The mind is a passive onlooker waiting for a chance to get into play. The mind is dependent on brain tissues, the neurons.

Secondly, remembering information. If there was nothing stored by brain cells, there is nothing to remember. I found the above definition of *memory* incomplete and not relevant to the current work on neuroscience. Looking for a more relevant source, I made use of DSM-IV, mental health disorder classification codes found in the psychiatric diagnostic and statistical manual for the diagnosis of brain diseases and mental disorders that I had used for several years. "Memory is a complex neuropsychological process involving both the neocortex and sub cortical structures of the brain. Memory involves the ability to learn new material, to recognize and register sensory input, to retain and store that information, and to retrieve or recall the stored information."[13] This definition partially conforms to my conceptualization of memory. It is defined as a process involving the neocortex and subcortical structures of the brain. The brain is the executioner of multiple actions resulting in habit and behavior among other things.

One simple request to you, where would you place the RNA messenger molecule originating on the DNA nucleus and travelling through the cell's plasma to be assembled into a protein at a ribosome? I know this molecule will be assisted along the way by transferring RNA among others, but it is still a long and perilous trip. This definition did not mention automatic motor skills, the genetic code, epigenetics, or environmental memory triggers. However, it is the most practical and relevant definition I have found,

My past experiences in psychiatric emergency rooms, private practice, and outpatient community clinics contributed to make me enjoy what I am doing now. It is not just addressing declarative and nondeclarative memories but looking for the genesis of survival tools my organism has learned to be able to celebrate my eighty-ninth birthday. Through evolution, my body has survived, among other things, by making use of learned and stored adaptive memories to identify, isolate, attack, and destroy invading pathogens. "This is a program of innate immune memory in health and disease. Host immune responses are classically divided into innate immune responses . . . and adaptive responses. The former reacts rapidly and non-specifically; the latter are slower to develop but are specific and build immunological memory. In recent years, emerging evidence has shown that after infection or vaccination, prototypical innate immune cells (such as monocytes, macrophages, or natural killers) display long-term changes in their functional programs."[14]

The author adds on the same page that mechanistic studies have demonstrated that trained immunity is based on epigenetic

---

13. Anthony L. Labruzza and Jose M. Mendez, *Using DSM-IV*, 121.

14. Netea et al., *Science* (April 22, 2016): 427.

reprogramming. I must clarify that epigenetics do not involve the basic gene code composed of the four nucleotides known as adenine, thymine, guanine, and cytosine. *Trained immunity* is defined as sustained changes in transcription programs and cell physiology. This line of study seems to be heading toward the genesis of the great macromolecules DNA-RNA acids that gave birth to all forms of life on this beautiful planet. Both forms of immune system are apparently tools developed by the union and cooperation of DNA-RNA and the mitochondria. These three great molecules are found within the same cell membrane.

There is no doubt in my brain that in evolutionary terms, my primordial cells are very old indeed. It is no exaggeration that my body ancestors' cells must have won many battles against myriad of pathogens. I must point out that many of these battles took places even before my neocortex was in place. There were battles won by electrochemical reactions, identification of toxins and avoidance, flight or protective shells, among others. There was not verbal abuse or sweet words to go the hippocampi to store and for us to retrieve. The hippocampi must have come later in evolution when our matrix cells divided and multiply into complex organisms. This complex organism needed a self-defense system as well as a memory formation and storage system to retrieve memories when needed. As these complex and useful systems developed, more sophisticated cells were organized into specialized groups of cells, forming our neocortex.

Most recently, after our long and old evolutionary trip, we have begun to explore the mysteries of our cells' wisdom. We have not only discovered our stem cells but amazingly embryonic stem cells as well within our stem cells using IPS. Induced pluripotent stem cells were pioneered by Shinya Yamanaka at Kyoto University in Japan. James Alexander Thomson, a biologist at Wisconsin University in Madison, is the first American to reprogram adult stem cells into IPS. In the beginning, this scientific discovery was thought to be a cure for many of our illness. Diseases such Parkinson's, multiple sclerosis, diabetes, spinal injuries, and many other health problems were thought to be cured by IPS. Although IPS research in America has slowed down, Japan, South Korea, and Singapore as well as some countries in Europe have reported successful application of IPS as a medical therapeutic tool. Interestingly and very curiously, *National Geographic*, July 2005, wrote: "With more and more countries aggressively developing stem cells therapies, the United States is in real danger of being left behind."

Well, it was not lack of brain power or human enthusiasm; it was politics and extreme conservative groups in America that derailed this technology in our country. Our government and extremely conservative organized groups should not intervene in a person's health if IPS research and its application as a tool in medicine saves life or helps patients with the above-named diseases. Sad but true, the federal government placed a stop to funding research on this promising new technology. Meanwhile, some countries in Europe and Asia, particularly Japan, China, and South Korea, have attracted scientists interested in IPS. Early in 2018, Japan took the lead and their health ministry gave doctors the go-ahead to graft sheets of tissues derived from IPS onto diseased human hearts. The team leader of this technology, Yosiki Sawa at Osaka University, clarified to concerned scientists around the world that the cells do not seem to integrate into the heart tissue. Instead, these cells release growth factors that help to regenerate the damaged muscle. "It is a very small beginning, three patients, but depending on their prognosis, another group of eight heart disease patients will benefit from IPS technology" (*Nature* 557 [May 2018]: 619).

I do not know if I should celebrate including macromolecules DNA-RNA and the mitochondria and stem cells in this memory voyage of mine. It scares me. The trip seems too long and endless in time and space. I feel somewhat relieved when I tell myself, "It is all in your brain. You do not have to leave dear Earth."

There are several thoughts in my brain urging me to stop, arguing I am too old. I calmed down these unpleasant voices by saying they are offensive to me. I have a good brain, and there are people who have dedicated a lifetime to this subject. I must go ahead without any more interruptions. Before I proceed with this self-imposed job, I must address some questions that are bugging me and trying to derail me from my objective. For instance, is learning and memory the same thing? If they are different, how? Which processes are used in each one? What part or parts of my body are engaged in learning, and what parts in memory, if separate? Besides cells, are my internal organs like heart, liver, and digestive system involved in learning and memory? Is moving my feet and sound making in my mouth immediately after birth considered a learning and memory domain?

"Learning is the process of acquiring new information, whereas memory refers to the persistence of learning in a state that can be revealed at a later time."[15] First, the author of this definition of *learning* called it a process. He wisely separated *learning* from *memory.* It is for me to find out when and where this process takes place. Who or what acquires new information? I am reading this book on cognitive neuroscience, and I know that I am learning something. I also know that in the process of learning, several groups of brain cells are engaged. However, an instigating thought inside my brain is telling me, "It is known that while a fetus is inside the mother's womb, it learned to make movements. Was the fetus acquiring new information while making movement inside the mother? Were these learned movements a response to our genetic code commands? Remember, it all began in the embryo with DNA-RNA from the mother and father."

Awesome! During inception, a process of life was begun without any conscious learning from either partner. Is it true? This is all genetics, but more than the four basic nucleotides *A, T, C, G.* There is epigenetics too. I do not want to say it, but I am timidly saying to myself, "Yes, I am genetically wired from conception." My parents were completely unaware, ignorant of the process of life taking place inside my mother's womb. I began to exist as an embryo without any notion of myself. Did my parents have a cognitive awareness that I was in a process of becoming another human being? How can I say that I was learning? Worse, how can the embryo, after nine months of gestation, say that he or she has memory of this learning process?

It all has been just simple chemical-electrical reactions of multiple basic organic elements. I must include here the birth of our great molecules, DNA-RNA. There is a question or mystery around these

---

15. M. S. Gazzaniga, *Cognitive Neuroscience*, 2nd ed., (New York: WW Norton), 302.

two molecules I dare not to write down here, but if I do not, I will be betraying myself. How did these two great molecules get the idea of building a wall or membrane around itself? Was the wall built to defend itself from an enemy? This type of thinking might provoke unnecessary but interesting and soul-searching criticism in which I am not prepared to get involved. I cannot and I should not ignore for a second my indebtedness to DNA-RNA great molecules if my search for memory, learning, and consequently, solutions to our physical and mental problems is to be of any benefit.

Chapter IX

Parkinson's Disease: My Experience

There are millions of Americans suffering from mental disorders and brain diseases like amyotrophic lateral sclerosis, muscular sclerosis, Krabbe, schizophrenia, major depression, bipolar, Alzheimer's, cancer, heart failures, and in a personal manner, Parkinson's disease. I noted above that we have begun to unravel the secrets of our matrix cells by induced pluripotent stem cells, creating embryonic cells. First, I placed my research interest in Parkinson's because it has afflicted my family and, second, because dopaminergic cells are involved in several diseases. Our symptoms of rigidity, tremors, shuffling gait, along with loss of smell, and as most recently discovered, death of neurons that serve the heart. These symptoms lead to pain and death. There are several therapeutic tools for Parkinson's such as levodopa, a precursor of dopamine. We also provide deep-brain stimulation by introducing electrodes at the thalamus to control motor symptoms. These are palliative remedies aimed at softening pain. I wonder how many of you reading this book can imagine how frustrating it is for me to press the wrong key and mess up the whole paragraph or page. How awful and painful it is when I go to a restaurant to enjoy my preferred bowl of soup and had to put it aside because my hands cannot deliver a spoon to my mouse. I end up asking for something solid to eat or the soup would be all over my shirt. I have been called clumsy, stupid, and a wild old country man.

Some people believe I can control the shaking of my hands and my body balance if I try hard. They argue that if I have faith and repeat some sort of mantra, my brain disease will disappear. They do not understand what it means to have dead cells in the brain. Thank God I am alive and keep working and doing my daily exercises. However, when I sit on a stool out in the park, I must wait for someone to help me get up. If I do it by myself, as I have done several times, I am afraid to end up flat on my face on the ground. Getting up from the ground when my cane is ten yards away is not funny.

"About 95% of people with Parkinson's gradually lose their sense of smell, but so do many people with Alzheimer and schizophrenia . . . Parkinson diagnosis can be made only once motor symptoms appear, by which time about half of a patient's dopamine neurons have already died. In Parkinson, we must learn more about the disease: the spectrum of symptoms presentation, clinical progression, *molecular changes* before and during the disease, and the effect of these changes on the brain *structure and function*."[16]

I am beginning to believe that there is light at the end of the tunnel. Researchers are talking about molecular changes *before and during* the disease. At a molecular level, we will be working on chemical signaling and attraction at the most elementary level of cell formation and behavior. This implicates gene-coding in protein processes. RNA, in its multiple forms but specifically, interference RNA and micro-RNA posing in different conformations, may be responsible for the death of dopaminergic cells. An adolescent's prefrontal cortex is susceptible—risk-prone—to schizophrenia when dopamine is implicated in this disease. Similarly, in Parkinson's, it is the striatum—substantia nigra pars compacta in the globus palidus—that is lost by death of its dopaminergic neurons. Loss of dopamine causes deficit in movement initiation and slowness of movement. Several past studies in patients with Parkinson's disease and animal models have shown that dopamine depletion leads to

---

16. *Nature* 538 (October 27, 2016): 4–7.

less vigorous movements but not ongoing movement. This could explain why individuals with Parkinson's disease tend to choose less vigorous movement initiation. Once on the move, these patients select to keep moving fast. It appears to us that they try to maintain a satisfactory balance between the motor cortex excitatory glutaminergic neurons and the dopamine neuron's deficit in the substantia nigra compacta. It seems as if it were a plan at the most primitive level of neuron and dopamine genesis. We can observe a person with Parkinson's disease initiate a movement with great difficulty, but once she/he moves, it seems a normal behavior.

Understandably, patients and their family want therapies for the illness now. Most are not willing to hear about promising research decades in the future. Consequently, scientists and pharmaceutical companies go for palliative remedies to help alleviate the pain. However, current literature on this subject, including that from renowned research institutions, are saying that it is imperative that we include the basic acids of our life in our investigations. The implication is that dopaminergic cell death or atrophy has an early beginning possibly at nucleotides level. There has been embryonic cells transplant done in vivo with mouse and lesser-complicated organisms. In the USA, however, this technology is at a partial halt. Some researchers have indicated that tracing dopaminergic cell death signals has been a hard nut to crack. However, the nut will never be cracked if research resources are not available or are restricted by its donors.

I remember J. D. Watson and F. Crick's arduous job searching for money or employment at a friendly institution where they could work on the structure of the DNA double helix. I am grateful and extremely happy they opened molecular biology to the world eager for these technologies. The RNA molecule and its single strand changing its conformation is surprising us in its many uses. Watson and Crick made researched biology the great discovery of the second half of the twentieth century. We in America are the center of research and development on this planet. We have the brain power and enthusiasm to open the door to nucleotide signals and reverse the tragic dopamine cell death. There are graduate students in this field who have not found their right place in research and are just holding a job. Similarly, some of our brilliant scientists have left for Southeast Asia where research facilities are open to them. IPS was first done in the Far East. Cloning is a European discovery, but South Korea and Singapore are leading the way to more discoveries.

We will have to work with our immune system in our nucleotide research. Are there kinase signaling deficits that prevent T-cells from entering action when there is a risk of premature cell death? These are electrochemical reactions for study at molecular level. At the morula and blastocyst level of embryonic development, we should be able to detect, anticipate, and make corrective measures to prevent disease-prone cells from maturing. We do have part of this technology at the present time. However, actual application of this technology seems to be slow in early diagnosis of diseases such as schizophrenia, Parkinson's, Alzheimer's, demyelination of neurons, and more. There is no doubt in my brain that some researchers are working hard to stop dopaminergic cell death as well as stopping cancer growth. Immunotherapies were the main subject of discussion at the American Society of Clinical Oncology meeting in Chicago in May 2014. In cancer research, the main approach seems to be how to stop growth. One of its tools is IL-2, a protein made by the body to spur the development of T-cells in response to threats such as pathogens or cancer. [17]

I said earlier that I could see light at the end of the tunnel. This group of researchers is not only working at molecular and protein levels, but they are also trying to reinforce our immune system to stop cancer from spreading. Advances are made in women's breast cancer by surgery as well colon cancer by eliminating the cancerous section of the digestive system. The horrifying fear is metastasis, cancer cells spreading, conquering, and destroying every cell, tissue, or bodily organ on its destructive path. Discovering a molecule or protein capable of connecting with cancer cells' signals and stop it will be a God-given blessing to all humanity and the team investigating it.

---

17. Heidi Ledford, *Nature* 509 (May 29, 2014): 541.

Philanthropies and universities must open their doors to talented, enthusiastic, and intelligent minority students to pursue a career in medicine and/or agriculture research for the benefit of people who will be the recipient of their discoveries. Whether we intend to go full-time researching the genesis of illness killing many of our brothers and sisters around the world or just maintaining our palliative effort, to lead the world, we cannot escape going into the cell's multiple vesicles and their signaling system. We began with DNA-RNA, nucleotides, molecules, and protein processes, building blocks of our entire body. I am not forgetting to mention research with neurons and glia cells implicated in our immune and nervous system. We are battling on many fronts.

We have had successes and setbacks, but we are moving forward steadily. We are moving away from the brain hippocampi and declarative memories to molecules, proteins, and cell vesicle signals. These are memories at different levels of communication and behavior. Generally, cells have a time frame of life from inception to death by old age. We may be able to identify markers indicating the lifetime for each group of cells in our body. For example, we may soon be tracing dopaminergic cells from the stage of morula or blastocyst to death at the striatum. Cells that die at the striatum must have some structural or chemical component different from cells that live longer. Tissue renewal is a fundamental process that relies on the regenerative capacity of long-living, self-renewing stem cells. The hematopoietic stem cells that maintain all blood cell lineage are, like other long-living stem cells, prone to accumulating DNA damage as they age."[18]

Working on young and old mice, researchers have found an increased abundance of proteins associated with inefficient DNA replication. The researchers explained that these excess proteins promote signaling by the enzyme ATR, which modifies many cellular functions. First, there is the macromolecule DNA having a replication problem. Secondly, there is an excess of proteins. Thirdly, the excess of proteins promotes signaling by an enzyme called ATR.

---

18. Bartex /Hodny, *Nature* 512 (September 14, 2014): 140–141.

Before I go to the next step, let's go back to the molecule DNA. From the nucleus of the DNA molecule comes the mold or template for protein and enzyme genesis. The template for protein synthesis is messenger RNA.

The fourth step is the enzyme ATR, which as a catalyst can modify many cellular functions. RNA is a single-strand molecule that during the translation process can provoke many serious problems. However, the DNA molecule commands most, if not all, processes in the cell. Despite the power of the DNA nucleus, an RNA strand, named interference RNA (iRNA), can prevent the formation of proteins and enzymes. This could lead to a confrontation between the most powerful molecules in our body. Can this be the causal factor for some genetic, chronic, and incurable diseases of our body? We have an immune system to detect and destroy foreign invaders, but this is an inside problem. Cancer has an inside origin. It is incurable except when surgery can do its job.

During the process of protein synthesis, the premessenger RNA is trimmed of debris before it gets into a ribosome. How can this be? There is an abundance of proteins implicated in the DNA replication. Are nucleotides forming the double helix affected by deficient DNA replication processes? If DNA strands cannot be replicated properly, what is happening at RNA level? Messenger RNA comes from DNA in the nucleus of the cell. However, it is the multiple forms of single RNA strands that do the job of hauling messenger RNA to ribosomes for protein synthesis. What appeared to be a simple cell-signaling problem has implicated the basic acids of life. I can understand how difficult and complex it is to track memory, electrochemical signals, dysfunctional proteins, and cells in our organism.

The matrix cells that made my body are old but wise and extremely sophisticated. In every step of my trip to Planet Earth, every physical, mental, and emotional progression was made to achieve a goal. To achieve this goal, protective measures were wisely executed. The living cell goal is to achieve maximum development by creating a self-protective mechanism like our immune system. In addition, it also has self-repair systems that add or delete genes, add or deactivate SNPs, swelling, sweating glands, pain signaling, and cleansing of our digestive system. All this wisdom exists in our cells. These are memories added to our brain through synapses or just simple chemical reactions. These original cells are very astute.

To confront outside predators, together they selected a group of cells capable of making decision for the whole body, the neocortex. Even the neocortex has specialized groups of cells known as neurons with a command post known as the prefrontal cortex. No wonder I have had so much difficulty tracking memory, signaling, electrochemical reaction among so many defense systems. Extremely wise cells of mine! Instead of adding weight to the body, these wise cells built a defense unlike any other in our solar system. Not even a T. Rex dinosaur has a chance of survival against a human being.

Aging, external threats, and internal dysfunction have left many scars that are threatening our health. Recently, we are trying to assist our organism's self-repair systems do its job. We are just waking up and looking at ourselves. How pitiful that we spent millenniums in the jungle coexisting with less-developed hominids. Science is holding on in our brain and is waiting to free us from our self-built prison.

Our neocortex took thousands upon thousands of years to develop into its current size and efficiency of functions. The brain was thought to be a mass of undifferentiated tissue. In other words, the existence of individual brain cells known to us as neurons did not exist. It was not until in 1905 that scientist Ramon Santiago y Cajal received a medal from the Berlin Royal Academy of Science for his discovery that the brain consists of individual cells known as neurons. Cajal also discovered that neurons communicate among themselves through synapses.

Our blood system feeds these brain cells with oxygen and glucose. Our heart pumps blood to our entire body. This wonder muscle of four chambers works without stopping day and night. When tissues in our body need renewal, the heart is ready to deliver blood, carrying self-renewing stem cells. "But during aging, stem cell function deteriorates. The hematopoietic stem cells (HSCs) that maintain all- blood lineages are . . . prone to accumulating DNA damage as they age. In the case of HSCs, the damage can reduce the cells ability to regenerate blood-cell lineages and increase the risk of diseases."[19] The bone marrow is the source of blood cells in our body. I should note that during the processes of replication and translation of both DNA and RNA respectively, there is a lot of stress going on adding to aging deterioration. I do not have to look anywhere else but myself. When I was celebrating my twenty-fifth birthday, I felt I could not ever age. I could walk, run, and jump for miles and miles without any pain or ache. When I was between eight to ten years old, I tried to fly by jumping from coconuts and mango trees. Now, age eighty-nine, getting up from my bed is a challenge. I imagine atoms, molecules and cells sharing and transferring among themselves electrons at different levels of signaling, provoking stress through oxidation to my body, among others. Deterioration of our cells through aging is undeniable. However, I keep asking myself how is it that some forms of life live longer than others?

There are trees that have overcome aging and live for thousands of years. Scientists have found cells thousands of years old buried deep in snow. Turtles live over a hundred years without visiting a medical doctor or taking over-the-counter medication, vitamins, and minerals to boost their body's multiple healing systems. I have seen people live over a hundred years surviving out in the country. They have relied on fresh fruits, vegetables, and home remedies for health maintenance. Our cells have survived thousands of years against external and internal enemies. I cannot but praise these wise cells to keep diseases at bay. They are sometimes called garbage collectors of our body, but never mind this unintended insult.

"Autophagy are key to health and long life. By performing self-eating- autophagy- they degrade or recycle toxic protein aggregates and cell components, keeping neurodegeneration and infection disease at bay and perhaps even control cancer."[20]

Yoshinori Ohsumi in Japan received the Nobel Prize in Physiology or Medicine in 2016. Doctor Ohsumi and others had shown that autophagy plays a crucial role in embryo development, cell differentiation; and

---

19. *Nature* 512 (August 14, 2014): 140–141.

20. Science 354 (November 7, 2016): 20.

immune response. *Autophagy* relates to the aging process and several other diseases. Although this discovery was done on nonhuman organisms, it will attract researchers, young and old, to take it up unto the next level, that of human cells. As I had said in earlier paragraphs, there is light at the end of the tunnel. There are autophagy researchers who praise the discovery and are waiting for funds to follow on Doctor Ohsumi's outstanding work and discovery.

Once more, cells belonging to our health defense system, be it enzymes or protein like kinases, T- and B-cells, can be strengthened and engineered in the fight against cancer cells and several other diseases. I do not hesitate to again mention induced pluripotent stem cell technology. While looking for researchers fighting bodily diseases, I came across protein P53. This seems to be another promising research approach at molecular and protein level. My immune system has very smart cells that have won many battles against internal and environmental threats against my body. Propping up enzymes, molecules, proteins, and cells in my defense system seems to be the most appropriate path to combat and win the battle for my health. I take OTC complements such as vitamins and minerals, which may have helped my system defeat or at least stop the invasion of pathogens to my health. I persist on continuing my fight against thalassemia and emphysema. These two diseases in my body remain undefeated. I am motivated to continue searching for an effective tool to become victorious against these diseases.

In thalassemia, there is a failure to produce enough red blood cells to carry oxygen to my lungs making me tire easily. This red-blood deficit has been a great concern in my family. The primary function of red blood cells is to transport oxygen from the lungs to the cells of the body. It also removes carbon dioxide from the body, transporting it to the lungs which will exhale it. Two of my brothers died from an advanced state of this disease. I wonder how environmental and internal immune system deficit as well as genetic risk factors became implicated in this disease. Can it be corrected or stopped at an early stage and thus boost my immune system at the bone marrow's hematopoietic cells? How about making appropriate changes at epigenetic implication level? Are short nucleotide polymorphisms involved in the deficit of hematopoietic cells ending up in the thalassemia disease? Can autophagic be engineered to recycle deformed cells and replace new embryonic cells from bone marrow cells?

I believe we have technology capable of pruning undesirable SNPs and rearranging risk codons. Some of this technology may have been done with lower animal experiments like mice, worms, and flies. Scientists may be waiting for funds to go to the next phase of research. The delay might well be that thalassemia is not a priority like cancer, Alzheimer's, schizophrenia, and Parkinson's diseases are. However, sickle cell is another red blood deficit or disease that could benefit from discoveries in thalassemia. These are genetically connected diseases that provoke pain, grief, and death to thousands of people around the world. A group of neuroscientists from the Max Planck Institute of Neurobiology in Germany wrote in November 2016 that the neuronal cell loss after brain injury or degenerative diseases currently cannot be repaired. A very sad but honestly stated truth from one of the most prestigious scientific research centers in the world. However, despite their honest opinion on neuronal cell loss, some researchers do not share the same opinion.

Susane Falkner's article on the journal *Nature* has shown that there is light at the end of the tunnel. "Using chronic in vivo two-photon imaging, we show that the embryonic neurons transplanted into the visual cortex of adult mouse mature into bona fide pyramidal cells . . . achieving adultlike densities of dendrites spines and axonal boutons within 4–8 weeks . . . After 23 months, the transplanted neurons are fully integrated, showing functional properties that are indistinguishable from their original counterpart."[21]

Yes, Peter, photons can be considered as a tiny particle of energy that makes up light. They may be considered waves as well as particles. We can argue that light is made up of small pockets of energy called photons. It is a subject I cannot deal with now. My enthusiasm and trust are that our scientific research groups around the world will not stop their work until there is a cure for degenerative and chronic diseases.

---

21. Susane Falkner et al., *Nature* 539 (November 10, 2016): 248–252.

I will show another disheartening but recent review on Alzheimer's disease. It is a brain disease that robs the victim of all its memories. A group of neuroscientists at the Massachusetts Institute of Technology wrote in November 2016 a realistic but disheartening research review on the most fearful of all dementia. "Alzheimer's disease is a progressive loss of memory and cognition, for which there is no cure . . . The insidious onset of AD-related memory loss has hampered progress towards effective therapies because cognitive symptoms emerge late in the progression of the disease . . . Despite extensive evidence to support amyloid-B, formation of neurodegeneration, drugs targeted at amyloid-B (amyloids are aggregation of proteins) have failed to reverse deficits in memory or halt cognitive decline . . . Preliminary studies using deep-brain stimulation electrodes in the basal forebrain cholinergic circuit and hypothalamus-hippocampus network have shown promise in slowing disease progression."[22]

22. R.G. Canter et al., *Nature* 539 (November 10, 2016): 187–194.

Chapter X

Genes and the Environment

It is scientifically proven that the environment plays a significant role on our genes' shape and function. For most of us, during the first twenty years of our life, we reach our maximum height, and the volume of brain cells has come to a relative stop, too. There is some brain growth in the hippocampus and ventricles, but pruning of dendrites may balance its growth. Our brain power is not dependent on the number of cells or weight but on learning, memory, neuron network, and experience. "Although brain maturation in all mammals relies in part on experience-driven development of neuronal circuits, human cognition depends particularly heavily on the experiential learning that occurs during our prolonged period of growth, which lasts up to two decades after birth."[23] Structure defines function, and all during this prolonged period of growth, our genes are busy processing peptides and proteins. Proteins are the building blocks of your body just as bricks are for building a house. However, there is a huge difference structurally and functionally in that proteins carry memory and messages for self-perpetuation. There is a continuous interplay between genes, protein, and learning. Thus neuronal communication and networking to produce an appropriate cognitive response. Some brain region matures earlier than others, with the prefrontal cortex the last one to mature. "Although it is the last

23. *Nature* 539 (November 10, 2016): 171.

region of the brain to mature, it is one of the largest sub regions of the human brain. The prefrontal cortex will claim for itself executive power and functions such as abstract thinking, creative problem solving, and the temporal sequence of behavior."[24]

Not a small prerogative of brain power considering it is the last region of our entire brain to achieve maturity, it seems the prefrontal cortex has accumulated and assimilated to itself the might and wisdom of all past experiences and memories of all other functional regions. I cannot but wonder what the frontal cortex will achieve in the future. We are consciously and purposely helping it grow cognitively beyond imagination. There are scientists and biologists talking about purifying, meaning getting rid of risk-prone genes and epigenetic intrusive short nucleotide polymorphisms in the human genome presently thought unnecessary. CRISPR-Cas9, recently discovered, promises to be a wonder tool to engineer our genome and clean it from degenerative diseases.

Most of brain cells known as glia cells used to be called junk genome. Some biologists used to claim that glia cells are remnants or leftover fragments of our genome that are unnecessary at present. They further claimed that we can replace obsolete and dysfunctional brain cells with induced pluripotent stem cells. I have not mentioned synthetically grown cells, but biologist and neuroscientist Craig Venter surprised the world with his synthetically formed bacteria cell in May 2010. During the last hundred years, we have learned more about ourselves and our solar system than our ancestor *Homo sapiens* did for the last seventy thousand years! We are leaving behind the old man whose past theories and beliefs kept him in a slavery state of his own making. Our brain is inviting us to open a door to its own cells and learn to use it as tool of investigation for self-growth and healing. For example, neuroplasticity is very important after we have had an injury in the spinal cord. A neural interface system enabled communication between the brain and injured spinal cord that paralyzed one leg of a

---

24. N. C. Andersen, *Brave New Brain* (New York: Oxford University Press, 2004), 68.

monkey allowed the monkey to walk freely despite the injury. In 2012, a paper described the first use by a *paralyzed woman* of a brain-controlled robotic arm.[25]

There was a time when we were taught that our brain at birth was a tabula rasa, a blank slate. We have made discoveries that most of our ancient teachers could not even dream of. We were held back by philosophers and religious leaders who claimed for themselves the truth and wisdom of the universe. Everything revolved around Planet Earth, and in the center was man. They could not understand or wish to learn that this ego-centric man was hindering his own growth. We are born with a set of attributes, both physical and mental, that are led by growth-driven cells to achieve a goal as if it were a predetermined plan. These are tools our matrix cells developed to protect themselves from inside and outside threats. We have been living in darkness arguing that we know everything that is permitted to us to know.

25. *Nature* 539 (November 10, 2016).

Chapter XI

Books on the Soul

We have millions upon millions of books describing the power and wisdom of our soul or spirit, yet we did not even know that we have a blood system carrying blood to every cell in our body. It was not until the seventeenth century that William Harvey showed us that there are arteries and veins in our body carrying blood cells to every tissue and cell in our organism. Even today, there are millions of people saying, "Do what your heart tells you." Does it mean we should ignore the blessing of having a wonderful organ of the body, the brain, which has saved us from beasts in the jungle, fires, hurricanes, and internal microbes? Yes, the brain has discovered medicine and tools to protect us from internal and external health threats. The brain has led us to repair another wonder organ of our body, the heart. It is interesting that ancient Egyptian used to take out the brain through the nostrils from the dead body for a trip to the afterlife. They used to save the internal organs of the body and placed them next to the body. The point is that we love to talk about things we are ignorant of while ignoring ourselves existing in the here and now. No wonder a college graduate now knows more of everything than most philosophers and theologians of times past. It seems to me that an ogre or beast exists inside each one of us, suppressing the wisdom of our nature-given brain. This ogre is abusing, exploiting, and placing into slavery the mentally gifted and nonviolent person.

Power and privileges come with the ability to destroy and kill. Learning and new skills were limited to the group or groups in power with very little share for the general population. In Europe, it was not until the Renaissance that merchants, traders, and individuals with a special skill began to demand a role in governing their society. However, it was not until the French Revolution of 1789 that men and women rebelled against the tyranny of the few who shared very little. Despite efforts to keep the human brain from expressing itself for the benefit of everyone in our society, there are men and women who sacrificed even their life in the pursuit of happiness and justice for all. Consequently, science is open to all, irrespective of gender, social status, religion, or race.

Despite the relatively poor prognosis on the cure of chronic degenerative diseases such as Alzheimer's, Parkinson's, amyotrophic lateral sclerosis, Krabbe, multiple sclerosis, schizophrenia, and a few others, researchers continue to explore the cell's secrets that have enabled it to survive millions of years. For hundreds of years, man has wondered what the human body is made of and how it is made. Up to last the two centuries, our Western civilization adopted and defended the religious belief that man was made from dust by an omniscient, omnipresent, and all-powerful deity we call God. Whatever happened to humanity was destined by the creator of heavens and Earth. The color of our skin, hair, height, luck, and even illness was attributed to the will of God.

However, there are people whose curiosity led them to question popular beliefs and practices. Among them was an Austrian monk and botanist named Gregor J. Mendel (1822–84) who questioned a very simple but interesting observation: why some peas were wrinkled and some were round. The difference between the pairs of peas he called heredity traits are what we now know as genetics. It was just *a simple curiosity placed to the test of science.* I can say that within each cell, there is a memory bank (genes) that have survived generation after generation through their expression in proteins (building blocks of our body) without losing its intrinsic properties. *I must repeat it, without losing its intrinsic properties.* Consequently, in families that tend to marry within themselves (cousins marrying cousins), you see heredity traits repeat over and over. In Mendel's case, he paired wrinkled and round for multiple generations, producing pure wrinkled and pure round at specific generations. This became known as dominant and recessive heredity traits or genes.

The point to remember is that the cell keeps within itself a treasure box full of survival tools for us to discover. One of these tools may well be the protein CRISPR-Cas system. Defense cell systems must have begun at prokaryotes (cells without a nucleus) to defend itself from virus and plasmid alike. "In addition, prokaryotes have evolved adaptive, heritable immune systems: cluster regularly interspaced palindromic (sequencing in reverse) repeats (CRISPR) and the CRISPR-associated proteins (CRISPR-Cas)." The potential application of CRISPR and its associated proteins as a research and therapeutic tool appears to be endless and extremely useful. It may open for us the cell defense systems against degenerative and chronic diseases such as Alzheimer's, Parkinson's, cancer, and many others. "The ability to easily program sequence-specific DNA targeting and cleavage by CRISPR-Cas components and CPFL, allows for the application of CRISPR-Cas components as highly effective tools for genetic engineering and gene regulation in a wide range of eukaryotes and prokaryotes." -[26]

26. Prarthana et al., *Science* 353, Issue 6299 (August 05, 2016).

Chapter XII

Prokaryotes and Eukaryotes cells

Millions of years ago, primitive cells called prokaryotes learned to defend themselves from external as well as internal threats. They evolved into more complex eukaryotes cells, and more sophisticated defense systems were developed. I cannot but look and wonder how this tiny living thing could overcome so many threats and obstacles and end up forming or creating a superior and brilliant brain like that of Albert Einstein, Stephen Hawkins, Pablo Picasso, and Mozart. Each tiny cell was able to establish a form of communication among the group that together came up with a brilliant brain like yours and mine. These cells learned to communicate by signaling each other through electrical and chemical processing. These wonder cells assembled themselves and changed their structure, functions, and environment. All these intriguing processes of change, evolution, intelligence, and growth exist in our body as an individual cell or group of cells like our brain. These cells were apparently dormant for many thousands of years.

Your parents and grandparents opened door to science so we can clean and make the necessary corrections to our genome and gene expression that are conducive to degenerative and chronic diseases. I am not exaggerating or engaged in delusional thinking. All the survival wisdom that I have mentioned is stored in our survival-driven cells. I already mentioned a few persons whose brain cells have made discoveries in medicine, agriculture, science, and biology for the benefit of mankind and abstract projection far ahead of our time. In medicine and biology, Marie Curie is a very good example. Ramon y Cajal discovered the neuron as the singular functional unit of the brain. Watson and Crick discovered the structure of the double helix. Gregor Mendel discovered genes. As recent as 350 years ago, discoveries were reserved for gods only. Today, millions of people around the world do not believe man landed on the moon. Not long ago, open-heart surgery was considered the work of the devil. Now, there are thousands upon thousands of honest and hardworking people trying to pry the door open so every man and woman who wants to make discoveries for the benefit of humanity can have access to science and its technologies.

As I pointed out earlier, within a cell, there are two basic acids of life: DNA and RNA. These two old nucleic acids hold within themselves the attributes of storing, transmitting, and expressing the hereditary traits—genes—Gregor Mendel had discovered at a monastery during the second half of the nineteenth century. This separate and distinct unit of inheritance places in an orderly fashion the amino acid sequence of polypeptides. Polypeptides are amino acids bonded together in a chain as in proteins. DNA provides direction for its own replication and, through RNA, controls protein synthesis. DNA is a two-strand structure while RNA is a single-strand structure. From the DNA nucleus comes out a single-strand structure known as messenger RNA, which is transported to a ribosome by a single strand known as RNA transference for protein synthesis. This long trip begins at the DNA nucleus and goes out of the nucleus through a nuclear pore in the membrane into cytoplasm. From the cell cytoplasm, messenger RNA is joined by other molecules on their way to a ribosome for protein synthesis. Although many commands and messages come out of the DNA nucleus, it is the old molecule-branded RNA that is responsible for much of the hard work within the cell. There is another molecule within the DNA membrane known as mitochondria, which is strictly a female contribution to genetics.

There are basic rules to remember about the genetic code. For example, DNA consists of four pairing letters—*A*, *T*, *C*, *G*— is double stranded, is a sugar (deoxyribose), and is found in the nucleus only. RNA is a single-stranded sugar-ribose with its base pairing of *A/U, G/C*. The RNA is found in the nucleus and cytoplasm as in messenger RNA. Transference RNA is found in the cytoplasm, and it assists in the translation of mRNA nucleotides code into a sequence of amino acids. T-RNA carries amino acids to the ribosome in a specific sequence for polypeptide synthesis. This T-RNA molecule must identify and recognize both the amino acid and the messenger RNA to be able to assemble it into a chain of peptides like our body's building blocks, the proteins. Note that adenine (A) always pairs with thymine (T) and guanine (G) pairs with cytosine (C) in the DNA double helix. In RNA, adenine pairs with uracil (U). The base or code thymine is not present in RNA.

During this essay, I will continue to move back and forth, connecting information that my brain placed together. I said together because now it seems to be associated to one another. For instance, the memory on Napoleon Bonaparte. That particular memory might have been the way I learned it and put it together for storage on my brain. The first stop to the final folder in the archives of long-term memory storage in my brain is the hippocampi. There is one in each brain hemisphere and is necessary for memory storage and recall of all memories. For the sake of simplicity, let's say I want to read about Napoleon Bonaparte, the nineteenth-century French military genius and emperor. His military conquest of Egypt, Europe, and his failure to capture Moscow opened my curiosity about the strength of this little but big military man. I will read every article and interesting biography about Napoleon. To get a better picture of his might, I will hang a picture of him on horseback. Most of this information will come to my brain through my eyes, making a stop at the brain region known as hippocampi. The image of Napoleon on his horse is so impressive to me that it takes first place in the mental image that I have made of him. I will carry this mental image of him wherever I go. This mental image will leave the hippocampi soon after my first reading and mental trip within. Napoleon's portraits, his picture on horseback, and possibly his meetings with his generals would go to the occipital lobe of my brain. When Napoleon crowned himself emperor, the limbic system of my brain would have been lighted like a Christmas tree of joy. Memory of dates and numbers of Napoleon's troops would be stored on my left hemisphere while memories related to time and space would stay in the right hemisphere. Proper names such as *Austria*, *Moscow*, *Rome*, *Paris*, *Cairo*, and *Madrid* would be sharing the frontal cortex, limbic system, and associate areas on both hemispheres. Most of those names and dates would be part of my working memory as I would be using it in conversation with friends and family members on daily basis. Most memory about Napoleon would be stored as long-memory because I would use it as a strong point of reference in history conversation. Similarly, I use the French Revolution of 1789 as a very strong point of reference to argue in favor of the power of the masses to bring about economic, social, and political changes in established systems. These memories of the year of 1789 would stimulate me to bring it into conversations with friends and relatives more frequently and passionately.

Therefore, synapses would be activated more frequently and intensively than other groups of brain cells not engaged in this memory. A big chunk of memories about Napoleon would be forgotten or hardly remembered because I did not pay much attention or invested enough emotion into it. For example, how many buildings did the retreating Russians left intact for Napoleon's troops' shelter? I did not care about it. Most probably, many details about Napoleon's defeat during his invasion of Russia hardly found space in my brain. Moreover, some memories may have touched a pain button inside my brain, and the process of repression took over.

How, where, and when did those memories about Napoleon become long-term memory, working memory, or forgotten memory are determined in some respect by my own past storage of memories? For example, did Napoleon's troops massacre or rape members of my ancestors? Equally important, did an ancestor of my family take part in defeating Napoleon's troops? There are multiple factors to consider before we assign a memory to a specific brain location. It is safe to say that most brain memory storage takes place in the whole brain. Consequently, a lesion, death, injury, or dysfunction of a group of brain cells engaged in memory storage and learning implicates the whole brain. We have shown you the sad case of H. M. whose hippocampi were surgically removed.

There is another sad case, that Phineas Gage (1823–1860) who suffered terrible injury to his brain while working for a railroad company. An iron rod destroyed his left frontal lobe. Phineas suffered profound personality changes. There have been other cases involving car accidents, gun injuries as well as in war veterans. These cases allow neuroscientists to investigate the brain's inner working, injuries, and diseases.

In learning and memory, we retain information that is necessary and essential for our daily chores and general living. We learn to wait for the green light or look on both directions before crossing the street because it is a necessary survival memory. Waiting for the green light or checking the traffic on both directions becomes a habit, a behavior, and you do it automatically. I have mentioned the hippocampi as the first stop for memory formation. However, under the current understanding of memory, the hippocampi are essential for acquiring new long-term memory but not for maintaining them. This hypothesis readily explains the symptoms of some individuals who suffer damage to the hippocampi. They cannot form any new lasting memories but can freely recall events from before their injury. Their lack of normal hippocampal function trap them in their past.[27]

I would like to remind you that the focus of long-term memory storage is centered at synaptic connections. These connections can change over time, depending on the activity level at each synapse. The frequency and intensity of synaptic connections is also implicated in toxic memories and repressed memories. E. Kandel cites Charles Gilbert of Rockefeller University who found that cells in the primary visual cortex undergo remarkable anatomical changes during learning. "We now know that expectations, attention, and previous images obtained through experience all influence the properties of these neurons by altering their synaptic connections".[28] The lesson in this quotation

---

27. Jane Reece et al., *Campbell Biology*, 9th ed. (Pearson, 2011), 1123.

28. Eric R Kandel, *The Age of Insight* (New York: Random House, 2012), 309.

has been established by Kandel in another of his books. There is an adage that says, use it or lose it. The discovery of altering synaptic connections seems to be the opposite to that. By reading, studying, and analyzing information, you can store what you have studied into long-term memory and become much wiser. You can appreciate this lesson at school. Children coming from a family with a tradition of book learning to provide their offspring with proper and relevant reading material normally do better in school than children from deprived or who are not interested in learning to improve themselves. Their children are not exposed to or motivated to improve themselves through educational tools. If a child or adolescent spends its free time just pushing a couple of buttons to see his/her hero knocking down his opponent, we can rest assured that intellectual synaptic connections will not be readily formed. Consequently, synapses with new information have not been activated to enrich it and improve its use for further learning. You can learn to ride a bicycle and do all types of acrobatic stunt to attract attention, but your chances of getting admitted to MIT or Harvard University will not be in your horizon. Studying new information is conducive to learning, and those are stored at synapses in your brain and my brain. A good thing to remember is that our brain uses our stored memories for problem solving.

Intuition does not come from nothing. Our brain has at least two hundred billion neurons with a potential of ten thousand synapses each. We have a brain capable of doing the unbelievable; it is up to you and me to make it a reality. It seems there is no limit as to how many memories we can store in our brain to fulfill our dreams. Our dreams reside in a memory that encourages us to go to the next step: research and lab testing. Memory is everything we are now and forever unless we suffer a brain disease, stroke, or accident. Every piece of information on which we focus our attention is working in a plastic or elastic fashion to make us feel happy. Many of my current memories are nurtured by my childhood and adolescent dreams and memories. Many neuronal circuits and pathways have been altered and, consequently, have enriched my brain during my eighty-nine years on this beautiful planet. It is not only my brain connections that made it possible but also connection with nature and particularly human connections. The information coming from my sensory organs—eyes, ears, nose, tongue, and skin—will be processed for delivery to the cerebral cortex. The exception is smell whose information is processed by the thalamus, which is an old group of cells in the center of our brain. If needed for our daily needs and survival, most of this information coming from our sensory organs will be used as learning tools at synaptic level and stored as memory in different regions of our brain. Some memories will become anxiety provoking, ending in mental disorders like PTSD.

Moreover, individuals with a genetic predisposition like in schizophrenia, bipolar and schizoid personality disorder, among other mental dysfunctions, may be in great risk of a psychotic breakdown. I am not referring here to brain diseases that are part of the public familiar conversation like Parkinson's, Alzheimer's, Huntington's, and amyotrophic lateral sclerosis. I am trying to separate mental illness secondary to a toxic memory or anxiety and pain-provoking episodes like rape, violent domestic assault, gang assault, overwhelming military stress combat from genetically predisposed brain diseases. Some of these mental disorders and brain diseases are the competency of specialized professionals like psychiatrists, neurologists, psychologists, and psychotherapists. In most cases, a medical doctor is needed to prescribe medication while in mental disorders psychotherapists or psychologist will oversee therapy along with a psychiatrist for evaluation and medication. Professional education and understanding of psychodynamics, behavior therapy, Gestalt psychology, or other cognitive and behavior-oriented therapy leading to a state license is needed for the treatment of mental disorders.

S. Freud, C. Jung, P. Pavlov, F. Perls, J. Watson, Donald Hebb, B. F. Skinner, Joseph LeDoux, and Nancy C. Andreasen have greatly influenced my professional career. Neuroscientist Nancy Andreasen wrote the following quotation in her book *Brave New Brain*. It stands out as a gold standard of my memories stored in my brain: "They have figured out how the hippocampus and the amygdala work together to produce built-in memories that may either help us or come back to haunt us, sometimes without our even being aware that this is happening."[29]

One thing I have never been able to comprehend is the strength and power of memories. I understand that it is not that I have a molecule built inside my brain from conception for an apple or a camel and recognize it after I see it on the real world. It is the strength of a toxic memory that I could not delete or forget about despite my efforts to do so. That memory blocked learning, social relationships and my desire to grow up a happy young man. I carry that painful and toxic memory wherever I went, even thousands of miles away from home. That toxic memory prevented me from socializing with young people while I craved for their acceptance and friendship. It kept me down, ostracized, and sad. I was not able to overcome the power of that memory and throw it away from my brain. The accusation by one of my teachers that I was not fit to learn and, consequently, go beyond elementary education had hijacked my neocortex. I was convinced my teacher's impression on me was correct. It meant, among several things, that I was a boy with below-normal intelligence. It reminds me of slavery and the Holocaust in Germany during Hitler's reign of terror. The teacher disqualified me from attending carpentry and electrical classes. He thought my future life was a hand labor at a dairy farm. My question continues to be, what is it that we encode in a memory? Eric Kandel has shown us the growth of new dendrites at synaptic level is the place for the storage of long-term memory. There was a toxic memory at a synapse that had been provoked by an insensitive teacher. It appears to me that I was stamped for recycle at a very sensitive phase of my physical, mental, and emotional development.

---

[29] Nancy C. Anderson, *Brave New Brain* (New York: Oxford University Press, 2004), 290.

Chapter XIII

Researchers have proven several times that an enriched environment promotes dendrite density, thus improving brain function. On the other hand, an impoverished environment can lead to shrinking dendrites, which means, among other things, poor memory and poor learning. Similarly, chronic stress can lead to destruction of dendrites, provoking poor memory and learning. Besides poor memory and learning, chronic stress is implicated in several brain diseases and mental disorders. PTSD has been well documented and proven that it is provoked by chronic stress. Chronic stress can significantly compromise our health and our lives by interfering with our immune system's functions. Chronic stress can either destroy or shrink dendrites in the hippocampi, thus preventing memory storage and retrieval. There is another very sensitive group of brain cells known as amygdalae that alerts us of imminent danger. When we encounter a life-threatening situation, the amygdalae neurons send signals directly to the thalamus and the autonomic nervous system (sympathetic section) for a quick reaction. These are automatic responses to stored memories of past experiences. This automatic response explains, in part, that once we lose our memory, we cannot respond to threats or to a touch of love. Perhaps, the worst-case presentation is severe Alzheimer's disease. The patient does not respond to any sensory stimuli. The patient's brain cells are blocked by accumulation of proteins. In addition, the neurons' multiple dendrites and its axon are entangled with each other, making it impossible to communicate with one another.

It does not mean that all stress is bad for our body. We need to be alert and on guard for self- protection from possible predators in big and busy cities, especially in poor lighted areas. Manageable stress increases alertness and performance. When we go for a job interview, we need to be on the alert and self-confident.

Do we encode lines of different shapes and sizes for, let's say, a camel that I had not seen until I visited a zoo in New York City? What did I encode from Napoleon's daring and genius military successes? What parts of his multiple memories were encoded in the limbic system, association areas, or the frontal cortex? I am assuming that I did not continue to grow or form more dendrites as per E. Kandel's experiments. Those neurons' selective swelling for long-term memory changes the structure of the host cell and, thus, its function. It would also be conducive to interference of space, function, and communication among neighboring neurons. If we only encode abstract symbols from lines and pieces of intercellular plasma and other junk material in the medium or even from glia cells, it may be consistent with our understanding that we are born and die with about the same number of brain cells. At my age, I can talk about memories when I was three years old. I can recall it at will, imagine those memories unconsciously repressed or simply forgotten but existing in my brain. We have the potential to store trillions of memories, but I doubt it could be done by adding physical synaptic sprouts at each neuron. We have been busy classifying and naming different types of memories, but how they are processed and stored needs further research and clarification.

Neuroscientists around the world are aware and concerned about how little we know about the working of the brain, brain diseases, and memory processing and storage. Craig Venter has voiced alarm on this issue. Mark Zuckerberg and his wife, Priscila Chan, from Facebook have set out a goal: curing, preventing, or managing all diseases by the end of the century.[30] Hopefully, they will attempt to bring together brain scientists from around the world. At present time, research institutions and countries are doing their thing almost secretly. When and what they share seems to be among trusted friends with a common interest. Worldwide basic research, including brain research, is slowly presented on the front pages of major scientific journals. Developing and poor countries gear their economies at providing a minimum of basic needs. Scientific research is not on the list of their population needs. However, the need and thirst for knowledge is there.

"For a country whose entire gross domestic product is just half of what U.S. government spends on research, Cuba punches above its weight in some areas of science . . . I think they have accomplishments they can be proud of when adjusted to GDP and the size of the country," says Michael Clegg, a geneticist at the University of California at Irvine.[31]

In the Western world, scientific research is shared piecemeal by a few countries like the USA, Britain, Canada, and in a lesser degree, Germany, France, Sweden, Italy, and Spain. Japan, South Korea, Singapore, and China are competing among themselves for leadership while making inroads in the West. There have been timid attempts both in the West and in in the Far East countries to come together in finding solutions to some degenerative chronic diseases like Parkinson's, Alzheimer's, multiple sclerosis, and schizophrenia. Countries in Eastern Europe and Russia seem to be ignored in worldwide research associations. Home remedies and over-the-counter supplements and medications are a multibillion dollars business in Western Europe and USA. In the USA alone, vitamins business has come to an $80 billion annual enterprise without much supervision from the FDA.[32] Vitamins are sold online, by local franchise corner store, and sophisticated beauty shops that combine beauty, health, and love in one very attractive package. There are radio and television shows and programs that appear to be partially sponsored by professionals who, by their presence, give credence to

30 *Nature* 537 (September 29, 2016): 595.

31. *Nature* 537 (September 29, 2016): 601.

32. *The New York Times* (May 27, 2016): B3.

the salesman claims on their product. Without adequate, efficient, and professional supervision, most shoppers do not know where the ingredients in the vitamins come from. Are they genetically modified? How do we know if the product does what the salesman claims is true? Has it been replicated by a reputable research group?

It is my health, my life that I am shopping for. I have checked out several scientific journals and prestigious educational institutions, including USA government health agencies, regarding vitamin and supplement claim on health. The verdict from among around twenty investigators is maybe yes for some individuals and maybe no for others. Therefore, I am back to square one in the search for a cure for my ailments. Many customers are turning to India and China for their medicinal approach and possible treatment. China, above all, is investing significantly in USA and England by sending young students to our most prestigious academic institutions. Hopefully, it will influence governments and educational institutions to soften their mistrust on each other for the benefit of humanity.

A case in point is multiple sclerosis that often strikes between the ages of twenty and forty when people are entering the workforce and raising families. MS is usually thought of as an immune-mediated disease in which misdirected T-cells, B-cells, and other inflammatory drivers infiltrate the brain and destroy the protective myelin coating around nerve fibers, disrupting signaling to and from the brain and spinal cord. There is a neurodegenerative component in progressive forms of MS that causes widespread destruction in the brain's white and grey matter.[33] The estimated number of MS victims worldwide ranges from 2.5 million to 3 million. At present time, there is not a cure for this disease, and the cause is basically unknown. There seems to be a connection between MS and vitamin D deficiency as it is more prevalent in people farther away from the equator.

Also, besides the vitamin connection, there appears to be a genetic and viral infection implicated in this hard-to-diagnose disease. In the absence of an effective treatment, MS patients have relied upon

33. *Nature* 540 (December 1, 2016): 2–9.

home remedies, over-the-counter medication, and food supplements to treat their symptoms. There are several drugs at present time aimed at relieving MS symptoms, but this is a progressive and degenerative illness, and therapies fail to work at different phases of progression. Recently, a group of concerned physicians and scientists are joining their expertise and effort to form a global organization for early diagnosis and treatment. Among them, we have the following groups of researchers testing drugs for symptoms: the French biotechnology company AB Science, Jeremy Chataway at University College, London, Robert Fox, medical director of the Cleveland Clinic's MS center in Ohio, U.S. National Institute, Tim Coetzee, chief advocacy and research officer at the U.S. National Multiple Sclerosis Society in New York City. Out of this groups of institution came a worldwide Progressive MS Alliance, a $24 million initiative teaming up researchers and advocacy nonprofit groups from around the world. Alan Thompson, dean of brain sciences at University College London said, "This has never happened in the world of MS before." [34]

My prefrontal cortex and limbic system celebrate my feelings of joy and success when I read new discoveries in this field. I pray for any significant benefits and have hopes for a cure for my past patients. Surprises appear on the horizon for the benefit of us all. The following discovery of memory retrieval from a mouse specimen makes my heart dance for joy and hope for victims and families of Alzheimer's disease: "Alzheimer disease is the most common cause of brain degeneration, and typically begins with impairment in cognitive functions . . . two pathological hallmarks seen in the late stages of AD: extracellular amyloid plaques and intracellular aggregates of TAU protein have been the focus of attention. In transgenic models of early AD, direct optogenetic activation of hippocampal memory engram cells result in memory retrieval even though these mice are amnestic in long-term memory tests when natural recall cues are used, revealing a retrieval rather than a storage impairment."[35]

---

34. Ibid.

35. *Nature* 531 (March 24, 2016): 508–510.

The authors of this work proposed strategies to engram circuits to support long-lasting improvements in cognitive functions. As I understand it, the issue has always been the recall and storage of memory, memories after hippocampi removal, injury, illness or accident to each hippocampus. H. M.'s removal by surgery of both hippocampi prevented him from remembering anything from conversations with friends and family members just few minutes past. He could follow you on a conversation if you did not change his train of thought. Suppose you met him at an ice cream parlor and both of you enjoyed talking about the flavor each one had chosen. On the way home, you may ask him if he would like to go back for another ice cream next week. H. M. would not remember his visit with you to the ice cream parlor. His hippocampi could not transfer that memory for storage in his brain. H. M.'s memories stored before surgery were available to him. What brain channels were available to H. M. to talk about his life history before surgery is not known.

In AD patients, neither hippocampus on both brain hemispheres has not been removed but are dysfunctional because of plaques and tau proteins. Moreover, there are no memories to be recalled because the whole brain has succumbed to the disease. Therefore, I am wondering where we should start the diagnosis for Alzheimer's disease. We all tend to forget things occasionally for multiple reasons: anxiety, marital problems, employment insecurity, depression, among many others. If there is no room for a scientific and logic-based diagnosis, medication would be uncalled for. What I am trying to imply is that unless we develop a national policy to bring genome analysis into everyday medical care, treatment is highly compromised. Early diagnosis of insidious, inherited, and degenerative illness is the first step toward healing.

There are several European countries that have started using genome analysis in some populations. In the USA, the armed forces have begun a partial project on limited bases. I recall enthusiastic but perhaps naïve pronouncements on accomplishment of cell biology, genome manipulation, and brain miraculous accomplishments during the 1980s. There is no doubt we have moved ahead in some areas, but there is still much to do in areas like brain degenerative diseases. For one thing, the cell's internal processes in most of the vesicles remain unconquered and unknown through our current research tools. We seem to be working slowly describing the neuron's body and its dendrites and axons. Despite short successes, neuronal chemical and electrical systems as relates to diseases are still in the unknown or incipient. There are theories about memory formation, but actual step-by-step of molecules' receptors for a memory presence anywhere in the brain is the dream of many cognitive neuroscientists.

Another area of research I hope we can pursue soon is the multiple jobs and power of proteins along their chemical and electrical signaling. We are an organism by and from proteins. And all cells are protein made. Somewhere in this process, cancer cells decided to disobey and revolt against the norm of the many that had reigned for millions of years. Their revolutionary signals are translated into metastasis, destroying cells all over the body. Our biology-oriented scientists and colleagues in the field of medicine have not been able to catch up to protein signaling and thus stop the revolution. Each human cell has so much information locked inside its membrane in the form of plasma and vesicles that we need a huge army of scientists to decipher and utilize its secrets. We have both the money and the human resources, but our priorities need to be redirected toward the good of all human beings.

It is very sad to learn that because of lack of funds, there are postgraduate scientists who are unable to pursue their career investigating the genesis of many of our most deadly diseases. Focusing our attention on prosthetic and drugs to ameliorate our pain is not solving our problems. Funds can be redirected toward basic research on our basic acids: DNA and RNA genes and their processes in the form of enzymes and proteins. Are we neglecting glia cells, histones, chromatin, and the whole of epigenetics input on the genesis of some of our degenerative diseases? At least I am finding relief at the attempt to establish worldwide associations to fund and share the information at a global level. Why can't gasoline- and diesel-consumer individuals, organizations, and countries demand that petroleum-producing countries contribute a fraction of their income to worldwide research? No single country can do it; we have had several empires, but all have collapsed.

What is even more important is we cannot do it by dropping atomic and hydrogen bombs on each other. Our neocortex has kept millions of past experiences in the form of memory at each one of our neurons and synapses. Our neocortex enhanced our primitive respond reflexes and electrochemical reactions into the most sophisticated and effective defense system known. There is no doubt in my brain that we will crack open the door to that old but responsive genetic system of defense. This is the century of memory exploration, synthesis, conformation, analysis, characterization, and application on multiple levels. I exist because my brain has stored millions of memories through my genes in the form of chemical reactions, unconsciously or from socialization after my birth. My very old brain, my brain stem, which consists of midbrain, pons, and medulla oblongata, is implicated with my eyesight, salivary gland secretion, bronchi constriction, slowing down of the heart's hyper-acceleration, stomach, intestine, pancreas, and gall bladder stimulation. Some of those systems in my body have begun to show symptoms of deterioration. My DNA/RNA acids and their protein synthesis did an excellent job for me to survive so many years of evolution while in the jungle. Survival in and out the jungle into large tribal groups needed larger and complex bundles of my brain cells, some becoming special and specific to my body's multiple organs. Among those distinct and special cells are my brain's neurons known as hippocampus implicated in learning and memory. I cannot fail to partially address another rather old bundle of brain cells known as thalamus deep inside my brave and efficient brain. It is the main input center for sensory information going to the cerebrum. Incoming information from all the senses is sorted in the thalamus and sent to the appropriate cerebral center for further processing.[36]

I have very good reason to be proud of so many implicit victories of my old but wise brain stem and thalamus to be able to survive my evolutionary trip at home, Planet Earth. Whether reflexes, chemical reactions, or plain simple declarative memories, I must celebrate my eighty-nine anniversaries successfully. Despite my inherited diseases,

---

36. Jane Reece et al., *Campbell Biology*, 9th ed.,(Pearson, 2011), 1115.

my brain, immune system, somatic cells, and nervous systems have sustained my body essential functions in an almost optimal condition. However, it does not give license to stop looking for the memory that keeps me conscious of my conscious self.

Before I move into another area of my brain, I must say something about the hypothalamus. It is located deep in my brain, very close to the thalamus. It is considered to be a very old group of cells by most neuroscientists. It cannot be placed along the frontal cortex or associate areas of my brain. But the hypothalamus, despite my efforts to find a place and time of its genesis, is like a thermostat and biological clock in a human body. Besides these two fundamental functions, it exerts control over another small oval endocrine gland, which releases several hormones involving the adrenal gland. The biological trinity composed by the hypothalamus, pituitary, and adrenal gland plays a very important and decisive role in human physiology and behavior. If I add the red-alert group of neurons known as amygdalae (one in each hemisphere), my body would be placed in a fight-or-flight state. My unconscious collective and current experiences with cobras and coral snakes, although plastic, provoke my heart to high speed with extreme high blood pressure. Under different circumstances, these unconscious memories would be an opportunity for me to pursue memory to its loci, but I feel compelled to move on without snakes.

Before I leave the brain stem alone during my memory trip, I must briefly touch on a communication system that goes from this old and primitive brain region to the neocortex. I am referring to the reticular formation. These antique fossil-like neurons form a formidable system distributed throughout the core of the brain stem that filters sensory input. It saves us from much irrelevant, offensive, noxious, and repetitive chatting and unnecessary information that comes to our nervous system. After relevant information is filtered, it is sent to the appropriate neocortex region. For my own curiosity and satisfaction, I look on crocodiles as brain-stem-behavior-driven reptiles without a limbic system or neocortex. They are obnoxiously aggressive and live to eat and reproduce only. What a poor and horrible creature to rule our planet. Dinosaurs were the ancestors of crocodiles. Nature took care of them around 68 million years ago. According to most scientists in the field, a meteorite fell on the Yucatan peninsula in Mexico and killed them all worldwide. The disappearance of these monsters makes me laugh and jump with great emotion.

Besides the frontal cortex input, for a visual recall of this memory, our emotions require activation of the limbic system and the amygdalae. Deep inside the brain, neurons such as those in ventral tegmental area, the nucleus accumbens, and other basal ganglia neurons would be releasing dopamine and endorphins, prompting you to dance even without music. Please don't mind me—I just wanted to celebrate the extinction of dinosaurs.

After briefly mentioning the prefrontal cortex I feel compelled to name some of the functions of our cerebral cortex. Cognition, awareness, memory, language and consciousness are core human functions residing in our cerebrum. Pierre Broca, a French physician, performed postmortem brain surgery on a patient who could not speak but could understand verbal language. He discovered that patients with this problem had damaged a small area in the left inferior frontal lobe. Similarly, Karl Wernicke, a German physician, found that patients who had damaged the posterior central area of the left temporal lobe spoke gibberish or meaningless speech. Thus the conclusion that the left temporal lobe is language dominant. Current scanning machines like fMRI have confirmed Broca's and Wernicke's findings. However, we must be careful not to compartmentalize our brain into fixed and specific jobs. For example, while you read these printed words without speaking it, you activate the brain's visual cortex, but if you read it aloud, the visual and Broca's area are activated. I must point out that both hemispheres of the brain work together, hopefully in a harmonious fashion for the benefit of our body.

The frontal cortex and specifically the prefrontal cortex make the final decision when conflicting thoughts are arguing over which direction to go. That will be true if your parents, teachers, religious leaders, or friends take your decision-making responsibility away from you. In psychoanalytic theory and practice, all the hypothetical persons I mentioned compose the superego. Consequently, if you are implicated in the above scenario, there is a strong possibility that you will store a conflicting and anxiety-provoking memory.

Anxiety and *memory* are nouns that most cognitive neuroscientists place it in the frontal cortex archives and some associate areas. Anxiety can be classified as a survival tool in humans. For example, you may feel anxiety when you are driving with your children in the car and you receive a radio alert of a coming tornado. A stress response in this hypothetical situation would provoke the release of two hormones from the adrenal glands, namely, adrenaline and cortisol. There is hardly anyone of us who have not had a bath of adrenaline during our life time. When I use the word *anxiety*, it means a subjective response to a past threatening experience. With fear, there is an object to be afraid of. For example, I am fishing in a public park, and suddenly, a wild bear shows in front of me. You bet I will be afraid. There are some individuals who must take tranquilizers before they get into an airplane. They may be afraid of altitude or believe the aircraft will crash before reaching its destination. Sometimes a human stress response provoking an alarming level anxiety may prevent the victim from defending himself from a predator and becomes immobilized in place and time.

If you are driving at night and get a flat tire in a very dangerous neighborhood and see three tough-looking men coming toward you, the flow of adrenaline may prepare you for a fight or flight. Situations like this happened to me in New York City. The first time, I was dragged out of my car and knocked down on the ground. Fortunately, police in an unmarked police car nearby saved me from harm. The second time, my car had a flat tire. It was about 11:00 PM. I saw two tall male youngsters and a heavyset girl coming in my direction. I became very confused and anxious. I could not open my car doors to fight or run away from what I anticipated was a threat to my life. In retrospect, it seems that adrenaline had hijacked my brain based primarily on my most recent experience. Luckily, the girl was the first one to say, "We are from the neighborhood safety patrol. If you need help with that flat, we can help."

I have indicated in above paragraphs that the amygdalae are our first guarding post, alerting us of danger ahead. It sends multiple signals to our brain and nervous system for appropriate action. However, we have consciousness and unconsciousness that will be implicated in our response behavior. Loss of situational control, toxic memories, unconscious negative memories, lack of social support, and hopelessness have a significant impact on our anxiety-provoking response. My response to the approaching girl is a good example. Emotionally loaded experiences leave emotional scars that are very difficult to overcome. Despite of my multiple good and bad experiences, my brain goes back to memory, central nervous system, neurons, axons, and synapses.

Chapter XIV

Long-Term Memory

For my own benefit and following the advice of some of my past teachers, I would like to go over my encounter with memory and its storage processing before I proceed to look for my original memory responsible for building my body. Through the printed words and drawings, I have formed a conceptual visual memory of how the world might have been million years ago when the Tyrannosaurus rex ruled Planet Earth. From the time of T. Rex, I move forward millions of years ahead to the advent of highly developed hominids beginning with Lucy some four million years ago. From Lucy, I tend to hump to a relatively close relative of mankind, the chimpanzee. I place him on around two million years back in our history. Leaving the chimp behind, I will mention the Neanderthal man who lived in Europe most of his time. Traveling eastward from Europe, we will find the Denisovan human in southern Siberia by the Altai mountains. Their DNA sequencing has shown them to have existed about 420 thousand years ago.

Using novel techniques to extract and study ancient DNA, researchers have sequenced an almost-complete mitochondria genome sequence of a 400,000-year-old representative of the genus *Homo* from Sima de los Cerros, a unique cave site in northern Spain, and found that it is related to the mitochondrial genome of Denisovan, extinct relatives of Neanderthals in Asia.[37]

Traveling back to our more recent era, I become selectively amnestic. I would love to stop in Greece and meet Socrates, Plato, and Aristotle, and Jesus in Jerusalem. I do not ignore Alexander the Macedonian, Buddha, Confucius, or the pharaohs of Egypt. Julius Cesar comes next in my mind followed by Attila, Genghis Khan, Queen Isabella of Spain, Columbus, Galileo Galilee, the Native Americans, and Napoleon. At this point of the trip of my historical memory storage, I abruptly halt at a towering figure in neurology, memory, and brain science: Santiago Ramon y Cajal. His lifetime work and discoveries on brain cell structures and functions is most outstanding and brilliant. Using staining as a working tool, he showed the medical and scientific world of the late 1800 and early 1900 that the brain is composed of individuals cells now known to us as neurons. Prior to his scientific description of brain cells, most people thought the brain was an amorphous mass of living tissue. Ramon y Cajal even proposed that connections between neurons he found in the brain are modified when memories are formed. The connections he was referring to are synapses.

"Donald Hebb, an outstanding Canadian psychologist of the twentieth century, advanced Ramon y Cajal's idea of synaptic alteration into a hypothetical rule of learning that would prove fruitful for scientists wishing to study the cellular basis of memory . . . the strengthening of synapses that are active during the learning process in turn forms the memory of what has been learned."[38]

Following memory formation, neuroscientist, Eric Kandel was awarded the Nobel Prize in the year 2000. He says that long-term memory storage requires the synthesis of new proteins. Kandel further added that in long-term sensitization, sensory neurons grow new connections that persist for as long as the memory is retained. E. Kandel

---

37. *Science Daily* (December 4, 2013).

38. E. Huang and C. F. Stevens, "The Matter of Mind: Molecular Control of Memory" in *Molecular Biology of the Brain*, ed. S. J. Higgins, vol. 33 (New Jersey: Princeton University Press, 1998).

added something new to Cajal's discovery of the neuron. He found that neurons grow new connections at synaptic level. These new connections by neurons can become a toxic memory like in PTSD.

Scientists, patients, students and the public have been greatly benefited from discoveries made by dedicated and enthusiastic researchers. They want to go to the core or the genesis of our degenerative diseases, and anxiety-provoking triggers. In general, medically oriented scientists want to go beyond palliative therapy into finding a cure to an illness. Charles S Sherrington (1857–1952), a British physiologist and Nobel Prize winner, strongly believed that our reflex responses made up of neuronal circuits are the building blocks of our nervous system. Much research has been done on the subject since his death in 1952.

Kevin J. Tracey, a neuroscientist from Long Island in New York, is a pioneer in bioelectronic medicine. His initial research focused on the body's natural reflexes as a matrix for our body's immune system. Tracey and his team were partially motivated by Broca's and Wernicke's speech localization, those discrete brain regions that control specific behavior, which led Tracey and his team to postulate that cutting the individual circuits connecting the brain and organs (in this case the vagus nerve) could reveal the identity of specific areas controlling TNF, meaning tumor necrosis factor. Tracey's intellectual enthusiasm, curiosity, and scientific acumen was noticed in a paper written by Linda Watkins from University of Colorado–Boulder. She administered a signal molecule called interleukin-1 that causes inflammation and fever. This experiment was done on rats. Working independently at Japan's Niigata School of Medicine, Akira Niijima also injected IL-1 in rats. Niijima discovered that interleukin-1 in animals spurred electrical activity in the vagus nerve traveling to the brain.

Tracey proposed that a simple reflex control mechanism would shut down inflammation and fever to minimize possible damage to tissues. Reflexes that tune the activity of different organs are also essential for regulating inflammatory reactions set off by the immune system. The vagus nerve, which receives and sends signal to many organs, plays an important role in regulating the inflammatory reflex. Recent research has found that implanting a medical device (like a vagus nerve simulator or a pacemaker for the heart) that stimulates a segment of this nerve turns off a key inflammatory molecule.[39]

Doctor Tracey continues to elaborate for us the progress made most recently. For example, Glaxo-Smith-Kline announced a total contribution of $150 million to begin charting a road map for the field. In addition, the National Institute of Health recently announced a $248 million program over seven years called SPARC. Bioelectronic medicine is of great interest to me in a personal way. My father-in-law, a brother-in-law, and CF, a very close friend of mine, a physiologist educated in Europe where he did most of his research, all developed Parkinson's disease. Most cases of Parkinson's disease are found in people above the age of sixty. However, about 10 percent of cases are hereditary forms with an early onset.

I took my friend, CF, to the hospital the day before his surgery for the implantation of electrodes deep in his brain and drove him to my home the next day after surgery. Another friend of mine had managed his grocery store for over thirty years before he became another victim of Parkinson's disease. In addition to Parkinson's, he lost the lease to his store and became very depressed. I was very involved with both victims of Parkinson's. They were intelligent and hardworking persons who lost everything at an early age.

This initiative by Doctor Tracey with money allotted by NIH, GSK, and other scientists in Europe and Asia is destined to include molecular and cell behavior and signaling in cancer and degenerative diseases. This approach will implicate very old brain nerve cells from the brain stem through the release of a neurotransmitter from the locus coeruleus and Ralph nucleus. The limbic system and frontal neocortex will also be included in this research. I should add that my interest in Doctor Tracey's research on the vagus nerve is of immense concern. I have suffered COPD for many years, and recently breathing is getting more difficult. My immune system, mucus, and inflammation are accelerating my departure from Planet Earth at least in flesh form. There is a limit at what current medication can do to stop this chronic

39. *Scientific American* (March 2015): 30–35.

disease. For this existential reason, I'm always checking out research advances in multiple branches of medicine and scientific discoveries that can be helpful to me, friends, and the public in general.

You might have already noticed that my focus in this essay is the brain and memory. Secondly, I have briefly gone into the body immune system, kinases, macrophages, T- and B-cells. It is amazing to me how the immune system responds to first-time illness and stores an adoptive response system (memory) for the second-time invader. As I see it, there are after-birth sensory memory storage, chemical memory, reflex-memory as well as electronic atoms sharing and molecules signaling and responses. This is broad, complex but very interesting research area to get involved in and study it with enthusiasm. It appears to me that there is nothing more interesting in the known universe than the study and mastering function and behavior of our human body. I would be very happy if during this memory trip of mine, I could find a tool to wipe out off my body all that hurt and provoke so much pain on me. My brain and my immune system would be placed on top of the list. It may sound a fantasy trip or fictional movie, but I am confident that not many years ahead, we will be successful. The biology that Watson and Crick left us is about a half century old. We have been working with proteins, synapses, genes, genomes, genome sequences, and many more. It has been a revolution in biology since the double helix occupied the front page in most science magazines and newspapers.

Chapter XV

Replacing Traumatic Memories

Scientists have been making great strides in how to delete, improve, and even create memories. Research published in *Cell-01-2014* revealed that drugs known as histone deacetylase inhibitors can enhance the brain ability to permanently replace old traumatic memories with new memories. Delving deeper, researchers from MIT have discovered a gene essential for memory extinction. published in *Neuron* (2013).[40] There are more recent researchers who have been working on memory formation and storage, mainly with our selective tool, the mouse. These researchers have had a decent partial success. They are strongly convinced that a breakthrough is coming soon. Worldwide research associations and local national projects that share similar concerns on human health are promising to tackle medical and scientific challenges for the benefit of mankind. Furthermore, philanthropies and private contributors are another incentive resources that facilitate science-motivated students to join our scientific workforce.

There is dire need to work with embryonic stem cells to replace dead cells in Parkinson's disease patients and brain and spinal injuries. Glia cells, the largest number of cells in our genome, although implicated in diseases and multiple functions of our brain, have not received adequate attention as an accomplice of some of our chronic and degenerative

40. *The Amazing Brain* (BBC, December 2016): 38–45.

diseases. Even cell signaling among our immune system like kinases, T- and B-cells, bone marrow, thymus, phages, and cancer metastases seem to be poorly understood. There is an abundance of enthusiastic human resources, but financial resources are miserably lagging. Ironically, the genome revolution of the second half of the twentieth century seems to have come to a halt. Patients with brain diseases like schizophrenia, major depression, bipolar, and anxiety disorders receive palliative medication that generally keeps them out of the hospital, but effective treatment aimed at the genesis of those illness are far away on the horizon. Sad to say, "there is no cure for schizophrenia that comprises about one percent of the world population. Major depressive disorder that affects one of every seven adults at some point, Alzheimer and Parkinson disease that disables millions worldwide have no cure at present time."[41]

A lasting cure expected by some researchers early in the 21st century is still in the waiting. The public funding promises by pharmaceuticals, federal agencies, and large research-oriented educational system and universities seem to have been forgotten by these organizations. There have been many genes identified in disease genesis or risk factors, but the next steps, medical translation, and clinical applications are far behind expectations. Whether it is lack of leadership or popular and scientific enthusiasm, it needs further exploration and clarification for the benefit of our health and country.

German doctor Alois Alzheimer was the first physician to describe the symptoms and pathology of dementia. He saw plaques and tangles of proteins now known as tau. Besides plaques and the protein tau, Alois Alzheimer also made drawings of microglia. This is a form of immune cells in the brain, resting near neurons. At the time, glia cells were considered junk cells, meaning noncoding cells from our past evolutionary history. However, neuroscientist Michael Heneka and his team have found that microglia have turned out to be central to the link between inflammation and neurodegeneration. According to some reliable sources, there is an estimated 50 million people worldwide with dementia. The Worldwide Health Organization projects it will rise to

---

41. Jane Reece et al., *Campbell Biology*, 9th edition, (Pearson, 2011), 1126–1127.

82 million by 2030. There are a lot of Alzheimer's disease patients and families out there waiting for a cure. This is a very tempting number of possible clients for pharmaceutical companies to think about.

However, there are skeptical professionals who believe that microglia are not the villain we should go after. Oleg Butovsky, a neuro-immunologist at Harvard Medical School in Boston, has said, "We just do not know enough about the biology yet." Similarly, neurologist Philip De Jager at Columbia University in New York City is working on an Alzheimer's therapy based on a microglial target but claims that T-cell from our immune system might turn out to be a very relevant factor to be seriously considered. As someone else had commented, we are just discovering the roles of glia cells. Most of our genome is made of glia cells. It might turn out that glia cells are responsible for most of our brain diseases and mental disorders.

Even our initial interest in memory formation and storage seems to have fallen far behind our expectations. My perception of the slow progress made in medicine with reference to chronic and degenerative diseases may be conditioned by naïve promises made by enthusiastic and well-intended scientists and scientific-journals reporters. They may have meant well, but did not anticipate the complex obstacles ahead before a solution could be found.

Perhaps a quotation from a research team dated December 2016 helps clarify my perception of current medical research: "Human embryonic stem cell has revolutionized medical research, thanks to the fact that they are pluripotent . . . However, embryonic development is substantially different in humans and mice regarding timing, morphology and at the molecular level. As such, comparison with mouse pluripotency might be misleading way to gauge (estimate or judge) the state of human cells."[42]

I specifically selected this researcher's experience because it illustrates the stumbling blocks that many researchers find when attempting to translate and apply research done in tools like mouse to human behavior and physiology. My heart was full of joy when the world's

---

42. *Nature* 540 (December 8, 2016): 211.

top research centers like Kyoto University in Japan and Wisconsin University at Madison announced their breakthrough discovery on pluripotent embryonic stem cells. I felt that my limbic system with all its emotion- producing groups of neurons had hijacked my logical, sound judgment and rational and intellectually reserved frontal cortex, providing or facilitating for my emotionally childish behavior to take over my personality. Foolishly, I had anticipated pluripotent stem cells as the magic tool to replace all dysfunctional and disease-forming cells in my body. I had gone in a delusional trip, hoping that soon that induced pluripotent cells was the golden bullet we were waiting for. I was celebrating it with professionals, friends, and even patients. I shed some tears at the time and accepted that a stumbling block can be translated into a good learning experience and memory.

My life experience of my early childhood and adolescence have left sour and toxic memories that haunted me for many years. I have tried hard to remove or at least modify and alter the strength of toxic memories still residing in my brain. Somewhere in a good book, I read and then applied a wise memory lesson: "Good and useful things do not come by easy." I see my brain doing so many wonderful tasks all by itself without any conscious effort on my part. My brain keeps track of everything I do to keep me alive and healthy. This is another reason to justify my unending search for the genesis of my memory archives and files. I follow new research on the brain to try not to do anything harmful to it.

Chapter XVI

Alois Alzheimer

It was Xmas evening when I was able to read the following on one of my favorite scientific journals: "The Antibody Aducanumab Reduces A/B Plaques in Alzheimer's Disease." The limbic system took over my entire body, and I reached out for a glass of wine to celebrate the good news. The introduction of the article reads as follows: 'The amyloid hypothesis posits that A/B-related toxicity is the primary cause of synaptic dysfunction and subsequent neurodegeneration that underlies the progression characteristic of AD . . . Treatment with aducanumab reduced brain A/B plaques as measured by forestair PET imaging in dose-and time-dependent fashion . . . The PRIME study shows that aducanumab penetrates the brain and decreases A/B in patients with AD in a time- and dose-dependent manner . . . the clinical and preclinical data support continued development of aducanumab as a disease-modifying treatment for AD.'"[43]

As far as I see it, there is a bright light ahead of us. Eric M. Reiman from the same journal wrote, "An antibody therapy markedly reduces aggregates of amyloid-B, the hallmark protein of Alzheimer's disease, and might slow cognitive decline in patients." Further research and replication will move this project ahead, hopefully putting an end to one of the most dreadful diseases to the human brain. To lose all memories

43. *Nature* 537 (September 1, 2016): 50–55.

accumulated since birth is like criminally stealing the identity of a person as a human being. It is reducing the victim to a bundle of flesh without meaning in life. We must put all our efforts and resources to stop this villain's assault on us. The dreadful fear in all of us is the loss of our mind. Taking away our memory storage is robbing us of our identity and ability to communicate with loved ones and friends. It is my fear that I will not be able to recognize my children, family, and neighbors. To think that I will not be able to be aware if I am clothed or naked at home or on the street is a frightening thought. I will not be able to learn and retain anything that is told to me' it is like committing suicide except that in Alzheimer's disease, you would not know what or how to commit suicide.

Treatment with the antibody aducanumab mentioned above has the potential to slow the decline of cognition, which involves, among other things, awareness, reasoning, bringing forward past experiences and memories as well as making rational judgments. During Alzheimer's disease, I will not be aware of threats posed by passing cars, fires, or poisonous snakes and wild animals. A bird would move away from me in fear of me hurting it, but I would not know if a wild lion or tiger is a threat to my life if I ever develop advanced Alzheimer's disease. This disease can affect anyone regardless of race, nationality, social status, religion, or philosophical orientation. From recorded history, some of the degenerative diseases I have mentioned have existed for millennia around the world. During Alzheimer's disease, my ability to learn, solve problems, write books and essays will be gone forever. I would not be aware that I ever wrote anything. It seems to me that I won't exist as a human being because my healthy brain will not be able to store and retrieve my past experiences in the form of memories somewhere in the synapses in neurons.

A ridiculous thought came to my mind. Suppose I transfer my memories to a robot. Would it be me or just a robot? Let's improve this hypothetical transferring of my memories by adding synthetic memory molecules? I can demand that only positive, intelligent, happy, and rational emotions be transferred to my robotic self. Before you use your brain to label this as delusions or science fiction, I love to say to you that it is not. Japan, South Korea, China, UK, and USA are already playing with these "funny" toys. They are primitive, simple, unsophisticated toys in a rudimentary stage of development. I recall how ridiculous it was to start a car in 1937 when a cousin of mine drove one at my father's home.

We humans can function and enjoy a relatively happy life with just one hemisphere of the brain. How about adding a synthetic one in the empty hemisphere? Perhaps we can drill a hole in the skull of an empty hemisphere and use embryonic stem cell lines to grow a new one. It will be cleaned of debris from our past primitive hominids. Surgeons do skull drilling for several reasons like in Parkinson's patients for deep-brain stimulation. Please do not accuse scientists of going against nature. Nature has been doing selective surgery and mutations for millions of years. We are far behind nature's wisdom to protect and guard our body from outside and inside risks and threats. We are just learning to investigate how our immune system communicates and performs its marvelous work.

A few days ago, I was outside in the backyard watching pigeons feeding from a small house I had built for them. There was a big bird, a falcon cruising the skies hundreds of yards away and above my head. I had a dove with a fractured leg in a cage in front of me outside the porch. I turned my back to get more food for the pigeon. As I turned, the falcon zoomed behind me, grabbed the dove by its talons, and stopped on the fence to eat it. The keen eyesight and incredible speed that the falcon used to snatch my dove left me speechless and thoughtless. That was nature at work. Highly developed eyesight is a survival tool of great magnitude. How genetics and environment complement each other to produce a superior eyesight or a brilliant human brain is a very tempting theme that I will put aside for the moment. The human brain's plasticity and human curiosity to explore the unknown is of immense interest today, perhaps more than ever before.

In retrospect, I should not be overly impressed by the falcon's highly developed eyesight to the point of undermining my own body's super defense system. For instance, when an antigen or harmful substance invades my body, my blood system develops an antibody to attack and destroy the invading agent. One part of my body defense system is the production of antibodies, which are very complex proteins that identify and bind to specific antigen and, consequently, provoke the immune system for an attack to destroy and remove the invading antigen. Basically, there are two cells engaged in the attack: B- and T-cells. Both cells have their genesis in the bone marrow but mature in different areas. When exposed to an antigen, these cells proliferate, and some of their daughter cells become memorylike cells. If the same antigen comes around to invade my body or your body, the memory cells recognize it and mount an immediate response. T-cells differentiate into several effector cells. Some destroy antigens directly while helper T-cells activate other T- and B-cells besides activating macrophages, a nonlymphocyte.

This brief introduction to my body's defensive mechanism has no reason to envy the falcon's keen eyesight. The immune system has evolved for many, many years overcoming myriad strifes and struggles to defend and protects us from many antigens ready to invade us. I have to say that I am alive thanks to my immune system and its memory cells. These cells immediately recognize past antigens and are joined by an army of T-cell helpers. Of course, there are few exceptions. One of these exceptions is when the defense system confuses pollen and certain foods as antigen and consequently releases an undesirable response. In this case, it is up to my brain to come up with an effective remedy. Among remedies against invading bacteria, our brains have developed a significant number of antibiotics to help destroy antigens. Before we discovered present-day antibiotics, home remedies such as garlic, onions, sulfur, turmeric, jalapenos, and many others were partially successful antibiotics. The body's response to a physical injury or an infection is generally accompanied by inflammation and fever. The assaulted cell or cells release histamine, which provokes or forces blood vessels to dilate; consequently, more blood flows to the assaulted area. During the inflammatory process, you will find the immune system's army of B- and T-cells along with macrophages carrying out their job, destroying the invading force.

Fever is another bodily response to keep antigens from further threatening our health. A blood test can reveal a lot of things about your immune system and general health. You have to have a blood test at least once a year. Our immune system of B- and T-cells, which are formed in the bone marrow, would later travel through our blood system. There is another fluid system called the lymphatic system, which is highly important to our health. T-cells are a type of white blood cells that circulate around our bodies looking for cellular abnormalities and infections. T-cells are essential for our organism survival. Although there are several kinds of T-cells, I would like to name two different types: killer T-cells and helper T-cells. A killer T-cell scans and can look inside the body of a cell and determine if a cell is infected. I can say that I have T-cells going around my body looking for and identifying invaders like bacteria. Killer T-cells travel scanning and destroying invaders while macrophages gobble up the debris. Helper T-cells are engaged in an immune response. The next time an invader comes around my body, immune system will recognize it. This recognition is a memory stored at synaptic level most likely in the form of a molecule. It is known as the adaptive immune system. When you get an infection, your white blood cells will be abnormally high.

Despite all the benefits we get from our immune system, there are serious problems too. Our immune system is responsible for rejection of transplanted organs, rheumatoid arthritis, multiple sclerosis, among a few I can name now. In early 1980s, for example, there was a dreadful disease. We could not identify its genesis. It was very scary to work in an emergency room drawing blood and caring for an HIV patient. I would say that although scientists have characterized the molecular markers of immune microglia, the brain's front line defense, how they function is not well understood. Without much evidence at hand, some researchers are suggesting that microglia are responsible for brain diseases such as Parkinson's, Alzheimer's, schizophrenia, autism, among others. Although there is a high proportion of genes linked to Parkinson's disease, schizophrenia, autism, multiple sclerosis is more highly expressed in microglia. It only suggests a link between microglia and several brain diseases.

We must not forget that glia cells have been the subject of several misinterpretations in the past. Brain scientists have made great advances identifying cells circuits and pathways implicated in several brain diseases, but individual cell research needs new initiatives. Perhaps using CRISPR-Cas9 to empower microglia will be another route to join the battle against the brain degenerative diseases and cancer. Our organism has developed a sophisticated and efficient defense system over the years; it is time for us to take over and give our immune system a significant push. Self-defense memories are embedded in our cells. Our next job is making use of CRISPR-Cas9 and its ally tools to strengthen these memories to meet new challenges. .

Please, I am not running away from my original quest or intent to find the genesis of my existential memory. I am sitting on a hard chair on the backyard of my house watching how a squirrel go around a cage to eat from a corn cob I had intentionally placed inside. The cage windows were closed except one that I had an opening of about a quarter of an inch. During the first trip, the squirrel went around the cage three times, trying to force itself into the cage through the narrow quarter-inch opening. I did not move from my chair and watched the squirrel leave the cage alone. I followed the squirrel walk on top of a wooden fence for about twenty yards. It had stopped twice, looking back once. During the third stop, the squirrel looked around and turned back to the cage in my porch. It went straight to the window with an opening. The squirrel did not care to look for another entrance. Somehow, the squirrel forced open the window and got in, except that the door had closed behind it. After enjoying some corn on the cob, the squirrel realized she was trapped inside. I opened the door, and the squirrel was quick to run away from me. Immediately after, I began to ask myself many questions about memory and the squirrel. Does the squirrel have a brain with neurons like mine? It was a female squirrel. It had a memory of the cage location in the porch. What forced the squirrel to come back to open the cage? Was it a basic need like food? Did it have babies to feed? I focused my attention on memory and asked myself, Was the memory of babies to feed the reason that forced it to take a second chance? The visualization of food in the cage and babies crying for food could have been strong stimuli to risk being trapped in the cage. One activating trigger came from its eyes, recognizing the corn was food. The second trigger came from the need to feed its babies, but from what part of the brain did it come from? Is it motherly instinct or love to account for this behavior? Is it a memory or just another biological need? Are there hormones involved in maternal love? If the answer is in the affirmative, when and how is it provoked? What other needs of mammalian species are wired on the brain?

According to some researchers, a long-term memory requires anatomical changes. Protein synthesis is implicated in the process of memory too. Where is the memory stored? Another group of researchers claim that a memory is stored at molecular level in the postsynaptic neuron. The message originates at the neuron's body as an electrical impulse, and it passes by the hillock (a kind of thermostat between a brain cell body and an axon) and goes down the cell's axon to its terminal buttons. This is an electrical and chemical process that causes or provokes the release of a chemical message at the synaptic space. There is electricity, and there is chemistry involved in delivering a message across the synapse. The big and seemingly unsolved question about memory remains partially unanswered: where is it stored? My logical and ever curious and research-oriented brain tells me that different parts of a memory are stored in cells in different region of the brain. I do not believe it necessarily has to be at molecular level, but where if not?

However, most neuroscientists believe memory is stored at synapses. Several brain and memory researchers have done excellent experiments replicating memory. Their experiments have shown that it is stored at synaptic level. However, almost all known research has been done on mice. Although the mouse it is considered one of the best tools to investigate, it is still a mouse and not a human brain. According to evolution theory, we separated from each other many million years ago. It appears to me that despite several claims on memory storage locations, we need better tools to approach this issue on human beings.

In memory formation and storage, electricity and chemistry are equally implicated. Although the message is a chemical one, the impulse to initiate and provoke the delivery is an electrical one. There are molecules at presynaptic and postsynaptic space, but if electricity is missing, there will be no message delivery. The same is true the other way around. The message comes in the form of matter from the presynaptic neuron toward the synaptic cleft. There are billions upon billions of atoms implicated in this process. Can bits of colors of a memory be stored at atomic level? For instance, Alzheimer's dementia; the flesh or matter is there, but there is no electricity circulating through neurons. Consequently, there is no trace of memory. All that exist there is entangled brain tissue and clusters of protein posing as debris.

Secondly, colors should not occupy space or weight. To colors, I should add beauty and sound as in music and feelings as in joy or pleasure. I am extremely hesitant to add time and most abstract thinking to atomic memory storage in the brain. Nanotechnology is doing it; why deny this attribute to our brain?

My personal experience seventy-five years ago was a toxic experience. I called it a toxic memory because of the pain it provoked in me. This is an attempt to find the storage loci for hundreds of bits of information that formed my toxic memory. I had the same teacher for three years in the seventh, eighth, and ninth grades. The class began at 8:30 AM. Before leaving for school, I had to help my father at home. There were goats, sheep, calves, turkeys, and even hogs to be fed before I left for school. Then I had to walk to miles on dirt roads to school. I had one pair of shoes just like everyone at home. I tried hard to clean my shoes and clothing before entering the teacher's classroom. He was a strong and tall man. For unknown reasons, he used to call me in front of the class of boys and girls to point out how dirty I, my shoes, and and pants were. He was known to be a tough and well-respected teacher by most people in the community.

I approached my father about my embarrassing situation at school, but his response was for me to hang on for another year. The teacher used to lift me up by my hand to make more fun of me. A couple of times, I thought of running away from home, but fear kept me there. The teacher made me repeat summer class twice if I wanted a ninth-grade diploma. The emotional damages that he provoked in me have never left me despite years of psychotherapy and training. The scars are there no matter how many miles separate us. I was able to alter, modify, or even delete part of his abusive behavior, but seventy-five years after, I remember his anger, laughter, and joy while shaking me in front of everyone in the classroom.

I left my home at age eighteen for the United States. I later travel to Europe. His face never left me alone. I grew up very insecure of myself, had stage fright, and avoided socialization in fear of rejection.

I believed his words that I was not worthy of a ninth-grade diploma. In the next paragraph, I will describe for you how I began to expel this abusive person from my brain.

I developed very useful tools to diminish, isolate, and destroy most of his power over me. Those memories are all over my body wherever there are neurons, including in my intestines. They are in molecules, atoms, and light signals in continuous communication with different regions of my neocortex. There is no specific location, no archive and file to look for all the bits of information that make up three years of actual mental and physical abuse. I used multiple tools to engage as many groups of brain cells to come into action and fight the invading intruder. I used hard tools such as rubber baseball bats to engage the motor cortex in the fight against the monster inside my brain. To engage my vision in the fight against him, I made use of drawing and graphics of his evil-looking face. In addition, I incorporated trusting powerful icons. On several occasions, I asked friends and colleagues who had similar experiences to be my co-therapists. We built a dummy in my aggressor's image with his name in capital letters placed on his chest. This revenge-looking therapeutic tool made me overcome the initial fear and challenge his cruel behavior. I whacked him with the baseball bat several times. I pulled his ears, eyes, hair, nose, and specifically hit his genitals. Castration was in my mind in this respect. Whenever I felt weak and depressed, I built a mental image of him showing my fear and anger over his abusive behavior. I wanted to take his macho brutal and punishing behavior away from him.

I must reveal to you that group therapy was very helpful in this respect. Having the approval and support of fellow trainees in sessions was a very useful tool to face and defeat the toxic memory of this monster.

Memories, specifically toxic memories, are by no means an abstract thought or concept. Doctor Eric Kandel and several other investigators have shown us that memory brings about anatomical changes. These changes are portrayed as brain tissue, not an abstract concept that pop ups spontaneously. Instead of sympathy, I am asking foe empathy. Take a few minutes and build a mental image of me, age fourteen to sixteen. Picture me being abused in front of girls about the same age as they are listening and watching me trembling and almost in tears. This humiliating episode took place at least three time a week for three years. If this destructive memory was built on brain molecular tissue only, it would have grown as a big tumor. I would like you to consider it seriously. There must be something else besides tissue growth in memory storage of long duration. PTSD in returning war veterans is a case for further investigation. Multiple studies have shown shrinking of the hippocampus's size in patients suffering from PTSD. This hippocampal change is suspected of causing an overvigilance in an individual, provoking inappropriate behavior and response. Symptoms of this brain disease are a consequence of the related overactivity in the sympathetic nervous system. Consequently, high levels of noradrenalin produce symptoms such as hypervigilance, exaggerated startle response, or tachycardia. A person with PTSD lives chronically with intrusive memories of perceived dangers.[44]

Researchers have used multiple scanners, but there is not a tumor or a quarter-inch tissue growth at a synapse. Another area with electricity and memory storage is, besides signaling among neurons, the translation and translocation of bits of memory to different brain regions. I believe that transferring bits of memory among groups of cells for storage, the messenger—meaning electricity—is also the message in the form of atoms. "From a physiological perspective, the brain must translate external sources of energy (sights, sounds, etc.) into electrical patterns the brain can understand. The brain then stores these patterns in separate areas."[45]

---

44. Nancy C. Anderson, *Brave New Brain* (New York: Oxford University Press, 2004), 308.

45. John M Medina, *Brain Rules* (Pear Press, 2014), 132.

As I said earlier, most neuroscientists believe that memories are formed and located at synaptic molecules. Meanwhile there are strong arguments that memory are stored in cells in different regions of the brain.

Very few researchers have gone as far as mentioning electrical energy as a place for memory storage. Making use of a teaching tool from my most favorite teacher, I like to repeat that atoms at each molecule implicated in a long-term memory or a toxic memory would be the right place to store feelings, colors, time, places, mental trips to another galaxy, or myself trying to domesticate a T. Rex. There have been several reports of neurosurgeons introducing electrodes deep in the brain and accidentally touching a brain fiber. Even though electricity was not used, the patient's behavior was affected. A response of tears, crying, and death ideation had even been verbalized by the patient in question. There are several interpretations and explanations for this type of behavior, which I dare not to explore extensively. However, a very thin electrode touched an electricity-producing cell or cells with the electricity-driven axon instantaneously activating millions of neurons to provoke an unanticipated behavior. It seems hard not to include atoms in the storage and retrieval of memories. Even more provoking, it produced tears and a death wish. Even if we accept the theory that memories are stored at specific cells, the cells can store anything because it runs by electricity. And without electricity running through neurons' dendrites and axons, there will be only tissue, inert tiny bundles of meat.

I need not to mention Alzheimer's disease once more; it is self-evident. I love to follow my professor's teaching advice that repeating a lesson reinforces a memory. Therefore, I will repeat that an axon is the part of the neuron that specializes in carrying a message away from the cell body to other cells. Most axons are covered by a fatty substance called myelin to protect and facilitate electrical signals to travel down the axon at much higher speed. The point I want to highlight here is electrical signals make it possible for brain cells to communicate among themselves. It is electricity that make brain cells come alive. Electricity makes it possible for billions of neurons to contact each other and formulate a plan of action as a group. This group of neurons can easily be the prefrontal cortex. The amygdalae could be another group of brain cells signaling the rest of the brain that there is danger ahead. Similarly, the thalamus is another group of neurons that makes decisions; it is a relay station of messages to appropriate neurons. There are situations when life and death is the decisive issue, and electrical signals must emerge to save the whole body.

Repeating a lesson sensitizes a neuron or groups of neurons, making it possible for a lasting long-term memory. I believe it is imperative that we learn that neurons employ electrical signals to send information from one part of the brain to another, from one neuron to another. During the following step, the neuron converts electrical signals to a chemical signal to pass the chemical message to another brain cell known to us as neurons. The receiving neuron converts the message to an electrical impulse, and millions upon millions of neurons and other brain cells participate in this wonderful process of my brain as well as yours. When a neuron is stimulated, an electrical impulse, known to brain scientists as an action potential, moves along the axon. The axon will be partially covered by myelin to allow electricity to travel faster without interruption. The myelin insulation prevents electricity from spreading or jumping to nearby neurons. The neuron is not a passive recipient of electrical and chemical input; it is a very complex functional unit of our nervous system. While the electrical impulse is traveling down the axon, there is a complicated balance of ions inside and outside the neuron that is constantly changing. Sodium, potassium, and chloride ion pumps make the brain cell's axon a most interesting part of our brain complexity and behavior.

Neuroscientists spend days after days working with and studying brain cells to understand our behavior, brain diseases, and much more. Much of the work is done in nonhuman brains. Neuroscientists have created even synthetic cells to advance our understanding of our human brain cells; however, they are still not human.

Chapter XVII

Chemical Compounds in my Brain

I have named several chemical compounds used by brain cells to achieve specific goals. They have different functions, are released from different parts of the brain, and often block each other to achieve their goals. Despite this complex and interesting interaction of chemistry and electricity, our brain is an organ of our body that has overcome myriads of challenges to make us master of this planet in a relatively short time. I will name a few neurotransmitters you are already familiar with while I was referring to several brain diseases.

1. Dopamine produces feeling of pleasure when released by the brain reward system. It has multiple functions depending on where in the brain it acts; it is usually inhibitory.
2. Glutamate is the most common excitatory neurotransmitter in the brain; it is important in learning and memory.
3. GABA (gamma-aminobutyric acid) is the major inhibitory neurotransmitter in the brain. It is important in producing sleep, reducing anxiety, and forming memories.
4. -Serotonin is involved in many functions including mood, appetite, and sensory perception. In the spinal cord, serotonin is inhibitory in pain pathways.

5. Norepinephrine acts as a neurotransmitter and a hormone. In the peripheral nervous system, it is part of the fight-or-flight response. In the brain, it acts as a neurotransmitter regulating blood pressure and calmness. Norepinephrine is usually excitatory, but it is inhibitory in a few areas of the brain.
6. Glycine is used mainly by neurons in the spinal cord. It probably always acts as an inhibitory neurotransmitter.[46]

Notice that half of these six neurotransmitters are implicated in forming and regulating memory. Dopamine, the feel-good chemical, is highly implicated in Parkinson's disease. Serotonin, besides the several functions already mentioned, is also implicated in depression. GABA and glutamate, both implicated in memory, do opposite work in the brain. One promotes action while the other suppresses the action.

We conclude that the basic signaling unit of the nervous system is the neuron. Our brain has about one hundred billion neurons plus many more auxiliary glia cells. Interaction among brain cells is conducted by electrical impulses provoking a chemical reaction or message. There is a space of communication between each neuron called a synapse. A message coming from a neuron through its axon is delivered at the synaptic space in a chemical form. At the other side of the synaptic space, there is a receptive molecule for the incoming chemical message. Therefore, you can see neurons communicating among themselves using electrical signals but delivering chemical messengers known as neurotransmitters. This messenger can stimulate or inhibit the activity of the receiving neuron. I named for you GABA and glutamate; one stimulates while the other inhibit actions. Basically, a neuron is composed of fingerlike branches extending from the neuron's body or center core; they carry information. Incoming messages will be processed at the center, also known as the cell's body. Besides these fingerlike components of a neuron, there is a longer cylinder-like extension from the body called an axon. It is insulated by myelin and carries out of the neuron's body a message. The axon is a very complicated section of the neuron. It is

[46] nih.gov, "Science Education," January 2, 2017.

in the axon that sodium, potassium, and chloride will be implicated in in the delivery of a message to the synaptic space.

When a researcher stimulates presynaptic neuron, there arises an electrical signal that evokes or provokes an action potential in the postsynaptic neuron. Electrical signals or information take place within a single neuron; however, communication between neurons in mammals, in most cases, is basically, a chemical process. Therefore, what crosses the synaptic space is a chemical message that binds to a receptive molecule in the postsynaptic cell. It is within the axon that you will come across with charged ions and molecules. Receptors are proteins that can bind to a specific chemical substance like dopamine. *A serotonin or glutamate receptor will not be receptive to dopamine.* This action or interaction between receptors and neurotransmitter is implicated in an action potential at postsynaptic cell. It may determine whether an action potential continuous its traveling toward the neuron's body. *Most neurons contain more than one neurotransmitter*, and some are excitatory while their counterpart are inhibitory.

I can imagine the neuron receiving hundreds of messages instantaneously, each at different locations in the cell membrane. Making it more interesting, each message may be responding to a different stimulus. For example, a stimulus from a pair of lovers and, in contrast, stress while taking the final chemistry examination. The memory that will be stored from this fictional scenario will contain many bits of information from different sources. It may end as a sweet ego-building memory or a sour one. I have had both experiences. Maybe my sweet and sour experiences could have been a positive stimulus during my search for memory.

Giving up is my nemesis, especially, when the goal is at hand. I have looked for memory storage in cells, molecules, atoms, and ions. For a moment, I suspected that it could be even in electrons, but all odds are against that. Luckily, I came across a very interesting article on learning and memory. Doctor Troy Littleton and his team of neuroscientists at Picower, MIT, wrote "Neuroscientists Reveal How the Brain Can Enhance Connections." It begins repeating the question I would like to answer, how the brain rewires itself in response to changing behavioral conditions—an ability known as plasticity. It behooves me personally and professionally. In the first instance, I was the victim of a teacher who punished me for being poorly dressed. It became a toxic memory that provoked much pain and implicated brain cells in a major way. My genes coded for the processing proteins that in time became parts of each synapse in my brain. How were these molecules processed while I was under duress by my teacher's abusive behavior? Is it going to shape the structures of individual groups of neurons? How was my frontal cortex and limbic system influenced or impacted while still in the process of cell maturity? How did it contribute to my cognitive and emotional health and development?

The author of this article says that it is important during early development but continues throughout life as the brain learns and forms new memories. Borrowing from the author's extensive experiences, he postulates that long-term potentiation occurs following persistent, high-frequency stimulation of the synapse. Keywords for me are *persistent, high-frequency stimulation of the synapse.* Imagine I had my torturing teacher five days a week, two hours a day. I feel fortunate I am reading and writing this essay today.

I praise my college professors and therapists whose understanding and empathy protected me from a nervous breakdown. The author of the article continues teaching all of us that his lab spent years on how the presynaptic cells release neurotransmitter in response to spikes of electrical activity known as action potentials. When the presynaptic neuron registers an influx of calcium ions carrying the electrical surge of the action potential, the vesicles spill their content outside the cell where they will bind to receptors on the postsynaptic neuron. Up to this point is the normal known behavior of synapses while a message is delivered. The researchers discovered, to their surprise, that at presynaptic level, they found that mini events were greatly enhanced well after the electrical stimulation had ended. This is novel and surprising—never ever had read anything like it. I would like you to apply this last bit of information to my teacher's experience. Consider the mini events following me after I left his classroom.

The research scientist describes the mini event for us in the following fashion. The enhancement of mini events appears to provoke the post synaptic neuron to release a signaling factor. *That goes back to the presynaptic cell and activates an enzyme called PKA.* This enzyme interacts with a vesicle protein called complexin, which normally clamps vesicles to prevent release of a neurotransmitter until is needed. Stimulation by PKA modifies complexin so that it releases its grip on the neurotransmitter vesicles, producing mini events. When these small packets of neurotransmitter are released at elevated rates, they help stimulation growth of new connections, known as boutons, between the presynaptic and postsynaptic neurons.[47]

Two main researchers at the above team involved in PKA and complexin study, namely, Littleton and Richard Cho, conducted this study at a type of synapse known as neuromuscular junctions in fruit flies. Cho added, "That machinery in the presynaptic terminal can be modified in a very acute manner to drive certain forms of plasticity . . . not only in development, but also in more mature states where synaptic changes can occur during behavioral processes like learning and memory." Applying it once more to my personal toxic experience, the overproduction of boutons between pre- and postsynaptic neurons may have a harmful effect that lasts for a lifetime.

From a strictly professional perspective, Maria Bykhovskaia, professor of neurology at Wayne State University School of Medicine, referring to Littleton and Cho's discovery, said that this "study is significant because it is among the first to reveal how presynaptic neurons contribute to plasticity . . . They used drosophila to determine the molecular pathway." Not intentionally repetitiously, but perhaps except for Eric Kandel, I had not come across any other in-depth study focused on the discovery of the mechanism of boutons sprouting at synaptic level and concentrating their research on presynaptic behavior. This is significantly important in learning, memory, and particularly, the mental health field. How good is repetitive drill learning? Is problem

47. MIT.edu/cms-2015-11-18, "Neuroscientists Reveal how the Brain Can Enhance Connections."

solving best learned by repetitive memorized lessons? How to spot a student acting out a behavior before it becomes a discipline problem?

I do not want to sound excessively alarmed by the sprouting and growth of unnecessary and harmful growing boutons in my brain, but I cannot ignore how it could be implicated in mental disorders and brain diseases like autism, schizophrenia, chronic anxiety disorder, depression, bipolar, and some personality disorders. I am referring to the pain, suffering, and discomfort provoked by unwanted anatomical modifications and structural changes in brain cells. It goes without having to repeat that structure determines function. If these anatomical changes take place at critical synapses in brain regions such as the hippocampus, prefrontal cortex, cognition and the emotions may be critically compromised. Also, I think of a child or student being bullied at school or around his home by neighborhood youngsters. I have read that in some cases, it has led the victim to commit suicide. It behooves school personnel and parents to keep an open eye to detect, identify, and take appropriate action and intervention to prevent a tragedy to take place.

My experience as a teacher and psychotherapist has placed me in direct contact with victims of abusive behavior by both children and adults. It has been very sad and painful to face both the victim and the victimizer. Not surprisingly, I have dealt with children abused by both parents. These are heartbreaking and frustrating experiences.

In retrospect, in the last few pages, I have written to demonstrate that the neuron is the functional unit of the brain. Also, that the space of communication between neurons is known as a synapse. Individual neurons use electrical signals to communicate internally and externally. Communication between two neurons is done through a chemical message delivered at a synapse. The message, for instance, glutamate, an excitatory neurotransmitter, binds to a receptive molecule at the postsynaptic neuron. There are excitatory and inhibitory neurotransmitters. GABA is inhibitory. Some neurotransmitters can act as both excitatory and inhibitory depending on their locations.

A neuron is basically composed of three parts: the core or body, dendrites, and axons. Dendrites receive information that will travel to the core; and axons send out information from the core. Axons use electricity within itself for communication. Axons are cylinder shaped covered by a fatty coat of myelin. The myelin facilitates speed and prevents contamination or spreading the electrical message to nearby cells. When a neuron is stimulated, an action potential, which is an electrical impulse, moves out of the neuron through the axon. At the end of the axon, there are boutons shaped like terminals that will fuse with the cell membrane and spill over a chemical message, a neurotransmitter. If the transmitter is glutamate, at the postsynaptic cell, a receptive glutamate molecule will respond. The neuron receives excitatory and inhibitory messages and is dependent on a balance and synchronicity of volume and strength of each of these messages. I learned also that there is neuronal communication in the opposite direction, meaning from postsynaptic cell to presynaptic neuron cell. This was done on mouse, one of the best research tools we have. Their organs are like those of humans. Eric Kandel had done similar work with a mollusk known as *Aplysia* several years ago. MIT seem to be the leading educational and research institutions focusing attention on memory location. I must temper my excitement, believing that I can point my finger to a specific cell or group of brain cells and say, "Please, doctor, inside this brain cell I have a toxic memory that needs to be removed by surgery." Not so fast, it needs to be replicated many times before it will be proper to try this research tool on human subjects.

I have been looking for my matrix memory in scientific journals and books with uncertain success. At times, I felt I was in the right direction, and the memory I am I am looking for was at hand. However, there is no reason to despair. There are many neuroscientists that have been working on memory formation and storage for many years. Learning that there are several forms of memory and that they are stored in different regions of the brain is a positive step forward. This learning experience moved me to investigate the neurons and their main functional components, dendrites, body, and axon. Stopping at the axon, I learned that it is insulated by myelin, and a deficit of this fatty substance may provoke several neurodegenerative brain diseases. Furthermore, I learned that an axon moves on electrical signals, and sodium, potassium, and chloride pumps are integral functions of axons. Ions are atoms or molecules with a net electric charge through loss or gain of electrons, and they move down the axon to deliver a chemical message at a synapse. Surprisingly, I learned that a synapse is where most all plasticity of the brain takes place. Neurotransmitters, the chemistry of our brain, implicated in our learning, memory, and behavior are delivered at a synapse to bind to a postsynaptic receptive molecule. With great joy and enthusiasm to pursue my goal, I learned that according to most distinguished neuroscientists, temporary memories are formed in the synapse.

The hippocampus is the master stop station for memory genesis. For the formation of long-term memory and storage, Professor Eric Kandel at Columbia University and most recently the Picower Institute at MIT have enlightened and elucidated for me how postsynaptic neurons are implicated in anatomical changes in memory genesis at synaptic level. What called my attention most is the novel approach and focus on postsynaptic contribution (retrograde) in molecular memory formation. "In addition to promoting synaptic growth, postsynaptic activity and Ca^{2+} influx through glutamate receptors triggers a 100-fold induction of presynaptic miniature release following high-frequency stimulation . . . These findings have revealed that drosophila embryonic NMJs undergo robust stimulation-dependent synaptic plasticity that is initiated by post synaptic Ca2+ influx and the release of retrograde signals that activate PKA and enhance presynaptic release."[48]

[48] Synaptic Plasticity, The Littleton Lab at MIT, Picower Institute.

Chapter XVIII

Ramachandran on Phantom Limbs

There are powerful discoveries that have the potential to strongly impact memory, learning, brain health, and diseases. It seems that the memory of my search is not that far away. It is almost at the tip of my fingers. All research points to my brain. My mind has just been stirred by a memory from neuroscientist Ramachandran's phantom limb research. People who have had a leg or an arm amputated most often feel the need to scratch the amputated organ as if it were still attached to them. In some instances, the amputated organ was felt even on the face of the victim. It seems the brain was rewiring itself. Is rewiring part of brain plasticity or another attribute of my brain? I need to investigate. Somehow, my brain was making associations I was not aware of. It is not a complaint against my brain. This is the way it has been working ever since I became aware of myself. The brain makes a map of our body, and we may not be aware of it. This source of energy may account for the aura some individuals claim they have seen around the heads and shoulders of some individuals. However, our central nervous system and peripheral nervous system can be the source of this energy. Is this aura, with its source of energy coming from our nervous system, the source of our soul if it exists? Can we group together matter, energy, aura, soul, and phantom limb under energy produced and released at brain cells? I wonder if scanners like fMRI and PET can be helpful in clarifying this apparently easy issue. Ramachandran and his team may have a response for us. My altered ego whispered to my ears, "Why not ask René Descartes? He had a good idea where the soul comes from and its location in our brain."

Peter cried out, "Please leave philosophy and religion out; stick to scientific research and discovery."

My main concern is finding solutions to brain diseases and mind disorders. I am interested in the electrical-chemical components of my brain. There is electricity, and there is chemistry, but my immediate question now is what is it that I am going to do with it? How can I use the power of electricity genesis in my brain to solve human problems? Electrical impulses are implicated in every memory that is formed or stored for further use and function. It is long known that a toxic memory can be deleted with a drug if administered immediately after the toxic event took place. Moreover, we have learned that there is postsynaptic retrograde response for lasting memory genesis by forming synaptic boutons or tissue sprouts.

Searching for my primordial memory is not just a human curiosity or, for that reason, at my age. It is brain stimulation for old and new neuronal connections to be activated. The hippocampi are two brain regions where new neurons are known to sprout spontaneously. Deep in the brain, there are two lacunae, ventricles, full of spinal cord fluid where new neurons grow. Therefore, my curiosity is a brain-stimulating exercise of great cognitive and learning value and benefit. I am sensitizing synapses repeatedly for the purpose of preventing or delaying brain degeneration and poor health. I do want to be able to take care of myself physically and mentally. There is no known better way to do it than mind and brain excitation and physical exercise. You may do it through chess, matching games, crossword puzzles, not-easy-to-solve problems, controversial philosophical or religious issues, or goalless meditation. Do something you have not done but wish you had. Have confidence you can do it even without any outside help. You will be surprised how many doors your neurons will open for you. Do not be afraid to go across. It is terra incognita for you, but billions of your neurons are waiting to surprise you. Do not allow a guilty feeling to keep you back.

Open the door to your unconscious self. Remember, everything we do or are aware of, our unconscious has already been there. We are only conscious of a tiny tip of our existence as a human being.

Neuroscience is helping us go deep in our brain to places and times we never suspected it existed. My brain, your brain has existed for thousands of years, but your time and my time have come to let the brain speak to us. It needs our help to clean it of degenerative diseases and other cellular debris so we may conquer new horizons of health and happiness.

Brain cell stimulation has been done directly by inserting tiny electrodes in monkeys' motor cortex. Plasticity of the brain has been demonstrated several times with victims of strokes, temporal lobe damage from car accidents, and hand amputation. Personally, I have witnessed hand and leg movement to almost total recovery. It is a demanding and strenuous effort by the victim to overcome pain, anxiety, and an occasional and unfortunate fall. Nevertheless, persistence and perseverance bring back much of the desired functions of the body organ.

I must not forget that embryonic stem cells therapy, either by cell implantation or any other technique, is another promising therapeutic tool at the tips of our fingers. I will never forget my favorite professor's last words to us: "It is in your hands to motivate and nurture future healers to use our method (scientific) to eliminate degenerative diseases from our society. I need, you need, we need support from the public, private institutions, and governments. Together, we shall overcome all human tragedies, maladies, and illness. We need to study, analyze, and scrutinize every neuron and groups of neurons to the minor detail of function and behavior. For instance, genes and environment, food and illness, behavior and protein formation, among many other subjects, are in dire need of researchers."

Areas in dire need of attention are depression and food intake, schizophrenia and vitamin and minerals deficiency, bipolar and lithium (salt), autism, genes and nutritional deficiency during pregnancy. We should get what our cells need in our body are able to divide and grow at each specific phase of development, or otherwise, the embryo and developing child may become a victim of an incurable disease. Cell specialization that are destined for your brain's prefrontal cortex do not wait until the father goes across the border looking for a job and sending home money for food.

We need elaborate basic lab work supported by worldwide scientific research instead of general, average samples for doctors to make educated guesses leading to diagnosis and treatment. In general, we, the public is not fully aware how the brain is loaded with unnecessary and poisonous memories that prevent cognitive and emotional growth and happiness. Take, for instance, lying to avoid an emotional setback and loss of a friend. Researchers at the University College of London reported how lying changes brain function. "They placed paid volunteers in a functional MRI scanner and measured activity levels of the brain's amygdalae. Money was used as a reward reinforcer of lying, which appeared to desensitize their emotional brain activity."[49]

I am taking a short break to read an interesting article related to one of my medical problems. "Both high cholesterol and diabetes can worsen cognitive abilities. Although statin drugs [I am taking simvastatin] are effective for lowering cholesterol levels and reducing stroke risk, not everyone knows that statins can increase the risk for diabetes, which in turn increases the risk for dementia."[50] Besides prescribed medication, there are other effective ways of reducing bad cholesterol level without risking your health. You must weigh the risk taking statins and its benefits. Sometimes the cholesterol is so bad and persistent that there is no other safe way to do it. In my case, diet, exercises like jugging, running in place, and some vitamins and minerals have been helpful bringing down bad cholesterol. We must be careful about our daily diet. For example, iron (Fe) and iodine (I), all forms of life need them. At my age, I tend to be deficient in red blood cells and B12. My digestive system does not extract enough from my food to maintain a healthy balance. It may provoke physical as well

49. Gary Small, *Mind Health Report* 9 (January 1, 2017).

50. Garry Small, *Mind Health Report* 9 (October 2016): 6.

as mental abnormalities. Another interesting piece of information I stumbled into is "that between 70% to 80% of the body's immune cells are in my guts."[51] There must be a healthy balance favoring good bacteria against bad bacteria for my immune system to prevail in any confrontation with those bad guys. My diet is heavy on fibers, including mangoes, vegetables, fruits, and a glass of wine. Avoid constipation and make sure you move your bowels every day. Drinking plenty of water, not any water but clean crystal-like water. High blood pressure, high bad cholesterol level, stress, poor diet, past genetic risk, and cigarette smoking are strong markers for some types of heart disease. Stress seems to follow us almost everywhere we go; it seems we live in constant fear and stress. Particularly, chronic stress is our greatest enemy today. I must remind everyone that chronic stress kills brain cells in vital areas like the hippocampus, a strategic center for learning and memory formation. The state of my health and my age forces me to look out for possible health risk and how to avoid it.

However, my trip on memory formation and its effects on my health and behavior has a long way to go. I have been dealing with neurons, axons, and synapses in the last few pages. Among the many questions that I have come across is, how do electrical signals do help making memories? Well, electrical signals leaving the hillock (a kind of thermostat between a brain cell body and an axon) goes down the very end of the axon to the synapse. The arrival of the electrical signal at the end of the axon causes the release of a message. This message can be in the form of an excitatory neurotransmitter like glutamate or acetylcholine. It diffuses across the synaptic cleft and binds to a receptive molecule of the adjacent receiving neuron. The adjacent receptive receiving neuron can receive hundreds and even thousands of these massages that may alter its behavior. Some messages are of excitatory nature while others are inhibitors. Drugs prescribed by medical doctors are meant to bind at specific sites of the brain cells when dealing with mental disorders and brain diseases. The prescribed drug can bring some balance if there is an excess or deficit of a neurotransmitter. If

51. David Brownstein, Center for Hollistic Medicine.

there is an excess of an excitatory chemical at the synaptic level, the drug may help by blocking receptors or reuptake at presynaptic level. A drug can be an antagonist or agonist, thus preventing or facilitating the release of neurotransmitters. In my case, I must take medication to help me breath because excess mucus accumulates in the lungs; it provokes a disease known as emphysema. I became interested in scientific journal articles trying to work at the gene level. The purpose was to intervene in the protein processing stage while on its way to the assembly post within the cell membrane called ribosomes. This approach may sound too far off with our present technology, but there are many curious minds eagerly waiting for a grant to give it a try. Most medications are of a palliative nature; they treat symptoms but do not a cure. That is one of the main reasons we must go back to the lab and do the research.

"Thou shall not stop your journey until you reach paradise here or in heavens," said a pastor at a local church. I am still looking for its meaning here. In the meantime, I will continue tracing memory in any form it presents itself. Emphysema has threatened to terminate my life by sending me to the emergency room a couple of times during Christmas season. My immune system is implicated in this plot, and breathing has been intolerable. Inflammation is an unwanted response to a health threat, often without a real enemy in sight now. There are B- and T-cells, lymphocytes, kinases, CD4, CD8 molecules, and macrophages involved in a fight that I did not provoke. At least, this is what I told the pulmonary physician.

Continuing with memory, memory T-cells are antigen-experienced cells that are generated during a prior infection. I have had several infections. T-cells, based on their memory of a past similar infection, can organize a swift and strong attack when they recognize a pathogen, an invader in our circulatory system. There have been several recent studies showing more attention to memory T-cells that permanently reside in the periphery with little or no representation in the wider circulation. A good conclusion is that memory T-cells have to be more effective at their own place of residence, the periphery. Where are my lungs considered to exist? In the wider circulation or peripheral?

Are there boundaries among different antigens fighting an invader, a pathogen? I must remind you that besides emphysema, I am a victim of genetically acquired disease thalassemia, which causes me to be poor in producing red blood cells.

Yes, my dear inquisitor, CD4 and CD8 are different types of white, yes, white blood cells, also called lymphocytes, which have markers on its surface. They are also known as T-cell helpers. When a virus or bacteria invades my body, they multiply in huge armies to destroy the enemy. My master inquisitor interrupted me once more and asked, "Why didn't those T-cell armies help you before you arrived in the emergency room?"

I promise I will find out later.

"Tissue-resident memory T-cells may be the immune cells of choice in designing strategies to stop pathogens before they establish a meaningful infection."[52] Lymphocytes are one of five kinds of white blood cells also known as leukocytes. All of them circulate in the blood. Commonly, we are more cognizant of B- and T-cells or lymphocytes. Both originate in the bone marrow except that T-cells mature in the thymus. Most T-cells in our body are divided into two distinct subtypes distinguished by the presence on their surface of one or the other of two glycoproteins designated CD4 OR CD8.[53] The word *glycoprotein* can be broken into its parts: *glyco* is a prefix used in biology, chemistry, and science. In general, it means sugar. Therefore, *glycoprotein* means "protein attached to a sugar." Carbon atoms are very versatile and exposed to multiple bonds. Mister Spock from *Star Trek* used to say that humans are made of carbon chips. Carbon isotopes are used to date fossils of dinosaurs and to find the age of very old trees and other disasters of long ago.

There are memories that seem to be pasted on my forehead. One of them is remembering messenger RNA coming out a DNA molecule and carried out by a transference RNA to a ribosome for protein synthesis.

52. F. R. Carbone and T. Gebhard, *Science* (October 13, 2014): 40–41

53. John W. Kimball, *Biology*.

And sugar! Oh, sugar, how much I love thee! It seems we all love sugar. The chemical formula of glucose is $C_6H_{12}O_6$. The white sugar we normally use to sweeten our coffee or tea, its chemical formula is normally different. At home, my parents always used dark sugar rich in vitamins and minerals. When we make a visit to a medical doctor, they love to take blood out of your body from your arm veins using a very thin needle. It reveals to the physician a lot of things taking place in your body.

In my case, red blood cells are low because of thalassemia disease. When I catch a virus or bacteria making me feel sick, there would be a lot of white cells, T- and B-cells, traveling inside my body. Among other symptoms, it would provoke inflammation, bodily pain and aches accompanied by feeling weak and possibly fatigued. At my age, I feel a lot worse. I feel very uncomfortable, anxious, often irritable, and angry at myself. I tend to become angry at myself because of Parkinson's symptoms such as tremors and loss of bodily balance. I knock down things on my way. Some individuals get upset when I stumble into them accidentally. Please, believe me. I want neither the disease nor the symptoms. It comes to my body without any previous invitation. I do not mean to distract you from finding out the outcome of my memory trip, but when I think that my ninetieth birthdays is not too far away, it is not a joke.

I strongly dislike viruses because once they invade my respiratory passages and lungs cells; the virus subverts the metabolism of the cell to make more viruses. At present time, there is no known medication against a virus, except to introduce a vaccine before you contract the virus. Viruses do their nasty job inside my cells in the safety of the cell membrane free from antibodies that might be present in my blood, lymph, or secretions. The job of destroying these terrorist viruses is left to my immune system. I do not like to write this down, but as my body celebrates another birthday, my whole organism is getting weaker. My hearing is in big trouble, my eyesight is trailing my acoustic attribute, my body balance is a disaster, my lungs are in desperate need of replacement, my brain has had two strokes, and my heart is sending me worrisome signals. However, I hope that my adaptive immune system (memory left behind after the first encounter with a pathogen) and my innate immune systems do not deteriorate like my other systems and can handle future pathogen invasions. I take selective vitamins and minerals followed by plenty of green vegetables and fruits. I avoid as much as I can riding subways and public buses where I will be exposed to uninvited foreign pathogens. I am not going to quit now, no, sir. Tomorrow is another day with changes that I will overcome.

Lung diseases provoked by cigarette smoking and underground mine toxins are known throughout the world. Illness and death by smoking can be stopped if we make it a serious national health issue. However, underground mining is an economic issue implicating millions of workers and government policies. I read about the number of deaths caused by coal toxins. In reading about mining, I also learned about collapses of tunnels burying hundreds of people in USA, China, Chile, Russia, Ukraine, and South Africa. Despite my concern about the safety of those people, I must focus on my lungs and emphysema.

Memory CD4 T-cells readily traffic through the circulation to provide protection at distal sites. Respiratory infection with influenza virus induces protective CD4 cells capable of migrating and establishing residency in the lung. However, it is not clear how and when CD4 memory cells in the lungs can be reinforced by innate or adaptive memory cells to mount an effective offensive against invading bacteria or viruses. I have had all vaccines during childhood, plus U.S. Army vaccinations and private immunizations during my lifetime. However, I become an easy prey for pathogens. I am a victim of chronic obstructive pulmonary disease, and pneumonia is a very frightening word that I try to avoid.

Following on CD4 and CD8, the influenza virus seems love to infect the epithelial cells that protect my respiratory tract. Neutrophils, the most common type of white blood cells, are the first immune cells to arrive at a site of infection. What a wonderful memory neutrophil cells have! These group of cells save millions of people from invading pathogens. They are phagocytic, meaning they can eat other cells but do not survive after ingesting a pathogen. A deficit of neutrophil may be congenital or acquired or derived from some type of anemia. It just happens that I am a thalassemia patient, meaning I am an anemic individual. Neutrophils travel in the bloodstream until they are called to the site of infection. Although neutrophils are known to recruit T-cells to infected sites during both bacterial and viral infections and in chronic inflammatory diseases, the molecular mechanism that links neutrophil and T-cell migration remain unknown.[54]

In addition to neutrophils, we have cytokines and chemokines, which are both small proteins secreted by cells of the immune system. Cytokines are basically considered small cell-signaling proteins secreted by many cells and are engaged in intercellular communication. Looking for a possible tool that could help identify and destroy pathogens that invade my body and attack my lungs, I am betting on CRISPR-Cas9 system. It has been described as a powerful genome-editing tool. It can be programmed to cleave specific DNA sequence by providing custom guide RNAs.[55]

Viruses invade the cell's protein-processing machinery and come anew in large numbers. There are scientists working with this relatively new research tool, and hopefully, a drug can be made to stop and kill the virus while inside the cell. Moreover, through GWAS cooperation, a computational system has been developed that "scores how strongly genetic variants affect RNA splicing, a critical step in gene expression whose disruption contributes to many diseases . . . among intronic [non-coding genes] variants that are more than 30 nucleotides [*A-T, C-G*] away from any splicing site, known disease variants altered splicing nine times as often as common variants."[56]

I hope you do not mind me going after research groups that appear to be closing in on technology leading to the genesis and cure of my diseases. Worldwide, there are millions of Parkinson's diseases, thalassemia, emphysema, brain, immune- and blood-related victims waiting for a relief from pain and praying for a cure.

---

54. Lim et al., *Science* (September 4, 2015): 1071.

55. Nishida et al., *Science* 353 (September 16, 2016): 1248.

56. H. Y. Xiong et al, *Science* 347 (January 9, 2015): 144.

I have gotten to a point where memories from the spoken word, whether conscious or unconscious, seem to intermingle with motor, visual, acoustic, smell, and somatic memories bordering on chemical reactions. Therefore, for the sake of my own sanity, I must go back and restore my memory trip. The most precious treasures I have had during my trip to this beautiful planet are the images, experiences, joy, love, failures, and victories over illness. In addition to these golden images, I treasure memories of lakes, flowers, birds, friends, and family stored in memory files in my brain. Out of the sum of all these stored memories done through an electrochemical system comes our sense of personal identity. It is our memories, in many ways, that make us laugh, cry, love, accept, reject, and above all, forgive and be forgiven. A memory can lead you to greatness or keep you down in failure. Holding on to a memory of a lifesaving icon can sensitize your neurons and overcome the most difficult problem. Positive thinking is holding to a creative success-oriented memory that allows no feeble thought to interfere. Memories can be the source of great pleasure as well as of pain in the manner of mental disorders like PTSD. A person without memories like one with the Alzheimer disease's is extremely painful just to think about. Just imagine that a loved one has no memory of any kind. Imagine sitting next to your mother feeding her, but she has no memory of who you are. You will have to feed, clothe, and bathe her because her capacity to store and recall memories is gone.

You must have a good idea of what memory is. You remember the date of your birthday as well as the names of your parents. You can easily remember the name of your first dog and the name of your closest friend during early childhood. Perhaps not as easy as your friend, you may recall your first-grade teacher. I recall my first-grade teacher's name. For Christmas, she gave me a T shirt with my name on it. So I will describe memory as the brain capacity of retaining most interesting dates, events, names, faces, colors, shapes, places, learning, and visual and emotional impressions that have impacted my life.

Memory implicates encoding, storing, retaining, and when necessary, recalling or remembering stored information from experience and past learning. It helps you build your own personality and self-identity as well as self-esteem. Through memory and learning, we build our own career or profession. If a person wants to become an electrician or plumber, he or she will be equipped with lots of memory qualified to do the job. Likewise, if a medical student wants to become a brain surgeon, he/she must have stored a lot of memory, information, and learning to be able to practice the profession. We know something about memory from our own experiences and reflection upon it. Scientists have learned about memory and learning from accidental medical cases like H. M.'s, car accidents, war veterans, and research, primarily with a mouse.

Next comes the question, Where is memory located at? Paleontologists have found DNA in humans over four hundred years old. Can we restore memories from these old men and women's acid of life?

As I mentioned in the above paragraphs, there are different type of memory, like short-term memory and long-term memory. Memory is first stored or processed in the hippocampus located in the brain's medial temporal lobe, which is an important part of the limbic system. There is one hippocampus in the left and right hemisphere of the brain. The hippocampus is mainly involved in temporary memory formation and storage. However, playing musical instruments and motor skills like riding a bicycle seem to be processed by unidentified loci at present time. Damage to the hippocampus can lead to serious loss of memory. The hippocampus may play a role in tasks involving difficult visual discrimination; in addition, the hippocampus is essential for recall but not for recognition.[57]

Paul King, a neuroscientist, says that the hippocampus is a temporary store for new memories, which are later transferred to the cerebral cortex, and is responsible for generating coding schemes. It has two unique physical properties: neurogenesis and highly recurrent circuits connections.[58]

Adding to the complexity of memory formation, storage and recall, there is the limbic system, which include, besides the hippocampus, the amygdala, cingulate gyrus, the thalamus, the hypothalamus, the

---

57. National Academy of Science-Medical Xpress-role hippocampus-10-2015.

58. quora.com. "What is the role of the hippocampus."

mammillary body, and other unidentified organs. All are involved in memory processing. I will need to study each one of the above-named groups of neurons if I want to be successful on my trip. The central amygdala is a forebrain structure vital for the acquisition and expression of conditioned fear responses and freezing behavior. As far as I remember, I have never had an accident that may have provoked an injury to my head. I have been good at remembering and storing information. However, my trip has been focused on finding my primordial memory, and of course, I have gotten into deeper territory implicating reflexes, chemical reactions, and molecule bonds. I must confess once more that electrical signals implicated in axon pumps (proteins) of ions of sodium, potassium, and chloride penetrate neuronal cell membranes but are not involved in memory storage; this remains inexplicable.

When I look out in the horizon and see so many colors, shapes, movements, sounds and feel the soft breeze coming from the ocean touching my skin, I cannot but argue that a spark of action potential (electrical signals) are triggering the formation of and recalling of memories. Axons without electrical signals would be inactive brain cables. For ions of sodium, potassium, and chloride not to be involved in the genesis of memory is highly questionable.

A relatively recent research on synaptic structure appears to shed more light in this interesting topic, nanocolumns at the heart of the synapse. A nanocolumn spans the synaptic cleft between neurons. Light microscopy and mathematical modelling were used to provide evidence for the existence of discrete protein-based nanocolumns connecting pre- and postsynaptic compartments.[59] It appears to me that this article demonstrates discrete physical tissue (nanocolumns), making it possible for retrograde communication for long-term and toxic memory formation.

Going after my primordial memory, my first learning experience has been exhausting, frustrating, and often irritating. It has frustrated me because my past experiences have led me to areas that I did not suspect I would be interested in. There are trivial memories like my

---

59. Tang et al. *Nature*. 536 (August 11, 2016): 151–210.

mother being an orphan at age five. What does it have to do with my birth and childhood experiences? Is it preventing me to open the door to a deeper inner self that is hiding to avoid further pain? Have I closed the door to the inner child that grew up begging unsuccessfully to a closed door to let him in? Are these made-up memories? Are these repressed memories of my childhood blocking my memory search? I have had several learning and emotional bruises and scars, but I have overcome most of it. I have had wonderful and lovely master teachers at a distance and in college who have guided me to a better world. I strongly believe that much of me is based on genetics, but genes can be modified by the love of people who live to enjoy the happiness they see in their neighbor.

Dear reader, I am not complaining. I am giving vent to my little one who has not given up. "Go forward, you are strong," he whispered to my ears. "You shall endure all challenges on your way. Deal with the present. The past is a done thing, and the future is for you to make."

I must continue my challenges whenever I find them. My last surprising news from memory and learning came from MIT Picower Institute's findings on retrograde synaptic formation involving protein processing. My understanding was that long-lasting synaptic changes demands gene transcription, which takes place at the cell nucleus's releasing messenger RNA to be transported to ribosomes, thus forming a protein. However, while I was going over one of E. Kandel's book, I found that Oswald Steward at University of California at Irvine had discovered that although, most of_*the synthesis of proteins take place at the body of the neuron*, some are done at the synapses themselves. However, to facilitate *sustained growth, proteins synthesized in the synapse are necessary.* That discovery was an eye opening for holders of curious brains. It seems Oswald Steward's findings are in consonance with E. Kandel's retrograde theory of synaptic sprouts on long-term memory formation.

Serotonin, a chemical and neurotransmitter, is made or processed in the brain and, primarily, in the intestinal tract. It is also implicated in the formation of long- term memory. E. Kandel found a molecule located at all the synapses of a neuron and is activated by the neurotransmitter serotonin, the transmitter that is required for the converting short- to long-term memory.[60] I had noted in previous paragraphs the significant role of chemical and electrical signals in the process of memory formation. Serotonin appears in our blood platelets and the central nervous system and influences many bodily and psychological functions. To my surprise, by association, serotonin brought to consciousness a friend of mine of long ago, René Descartes.

Descartes believed the pineal gland was the seat of the soul. We know the pineal gland synthesizes serotonin to produce melatonin that helps me fall asleep after a day's work looking for the elusive memory. A serotonin deficit has been implicated in depression, but depression has many culprit or origins. Major depression and bipolar depression both have a biological etiology.

I try hard to stay focused on a single argument or thought about memory genesis despite some interruptions. I cannot help it if thoughts of Descartes, Picasso, Kandel, Andreasen, C. Jung, or Einstein appear in my head. My style of meditation has two levels. The first one is to allow my mind to float freely without any attempt to stop at any interesting theme. I do not attempt to make any association of any kind. It may be frightening at times. Just stop if you feel overly frightened or scared. Do not forget that your brain works on electrical impulses. You have super alerting neurotransmitters like glutamate and aceticholinergic neurons that can fire by the millions. Some psychiatric medications are specifically prescribed to decrease or inhibit their release.

During the second level of meditation, I choose one that comes up spontaneously and brings along emotions and feelings. I make a pledge to myself not to discriminate on what is communicated. If it is noxious and I feel like throwing up, I do it without hesitation. I pay utmost attention to signals coming from my body as well as those coming from my brain. There are times when I have spent a week just watching a tiny avocado tree shoot out a leaf at a time. It is amazing watching and imaging a seed feeding the growing sprout hour by hour. It grows cell by cell using hydrogen, carbon, nitrogen under the shade of my home porch. The stem feeds from the avocado's round seed while its roots

60. E, Kandel, *In Search of Memory* (New York: Norton Co., 2006), 274.

grow down on the dirt. As the roots grow downward and the stem gets bigger and bigger, the avocado's round seed begins to shrink, degrades, and finally goes the same way all organic life ends up. One dies so another can live.

During these exercises or attempts to get my mind off any worldly nuisance coming from television, radio, telephone, or even neighbors, I disconnect it all. I must pay attention to my body messages to me. This might well be another unexplored pathway to my search of my primordial memory. My uninvited friend told me to watch out so I do not miss my ultimate goal—matrix memory—that could show up disguised as a glia brain cell. His observation was well taken. There has been ample documentation that we focus our attention in a determined shape, angle, object, or molecule and overlook the solution right in front of us. The role of glia cells and its multiple functions is still to be fully explored. When neuroscientists began sequencing the genome, they called glia cells just junk from past evolutionary phases. Although it seems to be an interesting and promising new territory to explore, glia cells are six to seven times more abundant than coding neurons.

For the sake of my mental health, I must decline my inquisitive friend's suggestion and stay in the safe side of my investigation—the neurons and learning in the process of protein and memory genesis. You may have several interesting suggestions helping me solve this issue, but I must decline.

A short initial setback is no reason or justification to turn my back and abandon my job. It goes against all my experience and the loving encouragement by my professors. In spite of his terminal illness, JLH's last words to me were, "Never give up, Jose. You have had difficult times, but from now on, the sky is opened to you." Also, I am reading a brief history of a medical doctor, Oliver Semler. He lost count of his fractures, but he suffered and endured twenty-seven surgeries. He is the victim of a debilitating genetic disease known as osteogenesis imperfecta in which brittle bones break easily. Sometimes, Doctor Semler has had a cast two or three times a year.

Stem cell therapy has been around for over eighty years. It probably began in Italy with many Parkinson's disease patients. However, full success has been elusive. This genetic disease must be worked on at the fetus/ embryonic level of development if permanent success is to be achieved. There are several obstacles to overcome before the treatment procedure is contemplated. Immune system rejection or inherited diseases from the mother are examples. Immune deficiency, like in my case, thalassemia is another impediment. Most past therapy has been done through the umbilical vein. It has been a very painful experience for the patient and parents. For researchers, from years of trial and frustration, helplessness has taken a toll, and patients have abandoned their dream of a cure. The next herculean attempt to get to the core of the disease will be prenatal stem cell therapy. The patient population will be pregnant women and their fetuses.[61]

Doctor Semler, despite his twenty-seven surgeries, is a pediatrician at the University of Cologne in Germany. He oversees over 200 children who suffer from the same medical condition as his. When I come across a man like Doctor Semler, I bow my head and beg the Lord to forgive me if the temptation of quitting ever crosses my mind. He is less than half my age, and he will dedicate his career, his life, and energy to find a permanent cure for a lethal genetic disease. In the meantime, he will do for osteogenesis imperfecta victims what other health professionals have done for him: "survive a lethal illness and enjoy his work and his life."

There is a lot to be learned from a project in which Doctor Semler and his team will be engaged for coming years. Hopefully, groups of experts in the field of immune systems behavior against pathogens and foreign substance from Europe and the United States will participate in preclinical trials. The past few trials ended in partial or total failure. As I have shown by work done at Columbia University in New York City and MIT in Massachusetts, there is a lot of territory to cover before we can say that we have mastered all cell functions and behavior. Researchers will be working at the very beginning of fetus and embryonic stem cell development and specialization. "The cross talk that occurs between cells of the innate and adaptive immune system and stem or progenitor

---

61. Jennifer C. Frankel, *Science* 352 (April 15, 2016): 284–287.

cells serves to emphasize the highly orchestrated fashion in which the human body conducts the daily business of survival."[62]

I hesitate to make an analogy between pre- and postsynaptic retrograde action potential and memory genesis in the presence of microenvironments determining cell fate. The above-quoted author alerts us that microenvironment influences more than stem and progenitor cells. Oliver Semler will be a first-class pioneer in such extrasensitive and mostly unexplored field of advanced medical technology and therapeutic tools. I hope that before embarking in this very delicate but necessary project, Doctor Semler and his team make connections with scientists in Japan, South Korea, Singapore, and China. Those four countries have more scientists doing research on or around the same objective than Europe, except Russia. Moreover, China wants its technology to be advanced to that of the USA and Europe. Therefore, they may be more inclined to pour more money into partnership projects involving USA and Europe. Doctor Semler's team has contacted some European scientists with minimal success.

The time has come for us to recognize that there is high quality research done on other parts of the world. There are multiple reasons for countries not sharing their discoveries even on projects aimed at the same diseases. Although there are ego and money boundaries that prevent closer cooperation, there are exceptions. For instance, a Gates-funded study headed by University of Oxford researchers and a similar effort by the World Health Organization individually developed worldwide standards for fetal growth. It involves a $29 million grant from the Bill & Melinda Gates Foundation in 2008. It is no small grant but is supposed to benefit everybody.

Europe and North America monopolized the field of psychiatry during the whole twentieth centuries. We are almost two decades into the twenty-first century, and there is widespread disagreement on DSM-IV as a reliable manual for the diagnostic and treatment manual for mental disorders and brain diseases. In many ways, psychiatry remains behind other branches of medicine. "Now we are in that 100 years ago phase

---

62. S. T. Badylak, *Science* 352 (April 15, 2016): 298.

in psychiatry, where we are just relying on people's description of how they are feeling as a proxy for what is going on their brain," says Prof. Craddock. [63] Doctor Craddock is a professor of psychiatry at Cardiff University in the United Kingdom. There are more critical reviews of the psychiatrist's manual known as DSM-IV than we care to publish in this essay. My personal experience is that it often does some unintentional harm because it is a very subjective diagnostic manual. I cannot call it a scientific research manual simply because there is no science involved in it. It is simply done on feelings and very subjective impressions by health professional, basically psychologists, psychiatrists, and clinical social workers. Occasionally, you will see a patient with incongruent main diagnosis of general anxiety disorder by one psychiatrist while another psychiatrist writes down "schizophrenic with persecutory delusions." Now, just imagine a nineteen-year-old youngster placed on high doses of haldoperidol, risperidone, or chlorpromazine. He will be a walking zombie unnecessarily exposed to death of millions of brain cells. Another psychiatrist wrote in the same article that DSM-IV, the most recent manual on mental health diagnostic manual, is a public relations disaster for psychiatry. And it is the best they could do. I have to say that despite the above critical professional's comments, there has been significant progress. Electroconvulsive therapy is still used as a therapeutic tool in some areas in the USA. Lobotomy, another psychiatric disaster, was used not very long ago.

Despite advanced sophisticated scanners like functional magnetic resonance imaging, widely known as fMRI, being used to observe brain activity closely, brain disease genesis remains enigmatic. Our treatment approach is basically palliative. Our genome continues to evade most scientific attempts to take it apart and reveal to us how it provokes degenerative diseases in our brain. It is no small task; there are billions of brain cells and trillions of synapses with which to work. Learning their signaling code is a big challenge ahead. We are just beginning to open a door to the processes of learning and memory formation. We are still deliberating about the basic roles of glia cells. Its role in memory

---

63. Jo Carlowe, "Sanity," *The Amazing Brain* (BBC January 13, 2017), 62.

formation and learning, the immune system, and degenerative diseases is almost entirely unknown.

The DNA double helix and RNA molecules are revealing surprising new roles to us. Once more, I must confess that I enjoy the trip that I have undertaken in this essay. It makes me very happy learning about new scientific and medical therapeutic discoveries aimed at bringing relief and cure for lethal diseases such as Alzheimer's, schizophrenia, cancer, Parkinson's, and other degenerative diseases.

After learning that I was admitted to a hospital for an emphysema attack, a friend of mine traveled three thousand miles to come to see me. He found me at home keeping track of tiny animals' and birds' behavior that I keep on my porch. He was not happy to see me outside my bed doing what I have been doing for a long time: keeping my brain busy. He suggested that I should stay away from furry animals. My friend argued that my pet rabbits posed a threat to my health. I, on the other hand, maintained that the love that I receive from my rabbits outweigh their threats. We solved our difference of opinion with a glass of wine. He helped me on my work for almost a week and left for home to work on a project like mine. My friend became interested in degenerative diseases when I showed him an international science journal with the following alarming information: "The number of dementia sufferers is set to explode to over 130 million by 2050, as more people survive into old age."

I have read several alarming news on the same issue in professional magazines and daily newspapers.

Chapter XIX

Genome Sequencing

Genome sequencing reveals the exact order in which nucleotide molecules are arranged along the strand of DNA. Each nucleotide molecule contains one of four bases: adenine (*A*), cytosine (*C*), guanine (*G*), and thymine (*T*). They are chemically attracted to each other, forming the double helix. A hydrogen atom binds each strand together. Some students liken it to the beads of a necklace. Nucleotides are the very basic acids of life. They are actively engaged in myriad of roles forming proteins for every organ of our body. There is not a single cell in our body that nucleotides are not involved. There are about 3.20 billion bases in a human genome sequence, arranged as complementary pairs that hold matching strands of the double helix together, and they are distributed across 23 pairs of chromosomes.[64]

To sequence a genome, the researcher must first break it into millions of bits. This was the original way used by the Human Genome Project, known as Sanger sequencing. However, neuroscientist Craig Venter used a different, faster, and a relatively inexpensive way for sequencing a human genome. Sequencing is the only way to uncover everything about the DNA that governs the genesis and progression of many degenerative, progressive, chronic, and hereditary diseases. Sequencing a genome has been made much easier, faster, and more

64. *Nature* 537 (September 8, 2016): 55–56.

accurate using the new technology known as CRISPR-Cas9. This science research and medicine tool has been the subject of controversy as well as hope for many people. Patients with a chronic and hard-to-cure disease are vulnerable to unproven promises from unofficial sources. We must always check out the source of information before we fall prey to unscrupulous individuals. Some are charlatans interested in fame and money.

Similarly, gene therapy was promised to be the silver bullet or panacea for all diseases a few years ago. It came to an almost complete stop after a hyper start was followed by several reverses in Europe and the United States. With pressure from chronic disease patients and their families to ease or stop their suffering, health practitioners overlooked several unintended diagnostic and clinical errors. There are several tools such as PET scan and fMRI that can assist health professional make an accurate diagnosis, but it is the human brain that makes the interpretations.

Medical scientists have been screening DNA for about half a century. Despite several success stories, close to 70 percent of patients with suspected genetic disorder fail to get answers. It has been said many times before that the hard part is interpreting the data. There is no doubt in my mind that we have made great discoveries in the field of medicine, but we are far away from the truth. There is a young lady who said, "After 30 years of waiting for a doctor's answer, I finally could learn the source of my problem, a mutation in a gene called GCHI." In another amusing medical news on November 27, 2018, scientist H. E. Jiankui from a university in the Shenzen province in China, made use of CRISPR-Cas9 technology for gene engineering in a human embryo. H. E. worked with a family of four; the father was HIV positive. The mother gave birth to two beautiful and healthy girls. Doctor H. E. Jiankui received worldwide condemnation for his unauthorized successful engineering. "I shall prove that I am not a clever speaker scientist in a way at all unless, indeed, by a 'clever speaker' they mean someone who speaks the truth" (Socrates in Plato's *The Apology*). Doctor Jiankui defended his unorthodox treatment with CRISPR-Cas9 at a Hong Kong meeting full of skeptical scientists and media.

In addition, "an international team of researchers has used CRISPR-Cas9 gene editing to correct a disease-causing mutation in dozens of viable human embryos. The researchers targeted a mutation that causes the heart muscle to thicken; it is the sudden death of young athletes."[65]

It is known by many scientists that China has already used CRISPR-Cas9 to alter disease-related genes in human embryos. Sweden and the United Kingdom have used the technology to study the early stage of a human embryo development. In the United States, scientist Shoukhrat Mitalipov at the Oregon Health and Science University at Portland, did embryonic experiments. As of 2014, the United States government does not allow federal money to be used on embryo research. "Dr. Mitalipov did not take the typical approach of inserting DNA encoding CRISPR components into cells. Instead, his team preferred to inject the Cas9 protein itself, bound to its guide RNA, directly into the cells."65

There are several diseases for which we have not found a cure despite outstanding, well-trained, and enthusiastic scientists doing research in well-equipped laboratories. There are thousands of people in the world dying of cancer. However, the cure seems to be limited to surgery during the first stage of growth and is limited to colon cancer and breast cancer. The promising and worldwide appreciated technology known as CRISPR-Cas9 seems to be held back by popular opinion and politicians. This new technology can add, alter, and remove parts of our genome suspected of serious health problem. It has been used on nonviable human embryonic stem cells. The result has been very satisfactory, but it has not been taken to the next step. In the past, some scientists argued that the need to use viral vectors has slowed down research and clinical practice. However, in July 2018, a research team has published in a science journal that they have developed "a CRISPR-Cas9 genome-targeting system that does not require viral vectors, allowing rapid and efficient insertion of large DNA sequences."[66]

Another research team reported that they used CRISPR-Cas9 to identify genes that modulate the response of cancer cells to methotrexate,

---

65. *Nature* 548 (August 3, 2017): 13-14.

66. *Nature* 559 (July 19, 2018):.405–406.

a chemotherapeutic drug. I hope this most recent use of CRISPR technology can be applied in the immune system to identify and destroy cancerous cells. Our innate and adaptive immune system has a history of proven powerful, experienced, and effective antigen killers, T-cells. CRISPR could be used to alter and reinforce T-cells to go after cancer cells and destroy them. According to MIT technology review from April 2019, CRISPR-Cas9 has been used in treatment for cancer patients.

In addition to the work done by the above researchers, a molecular biologist at the Salk Institute for Biological Studies at La Jolla, California, used CRISPR-Cas9 to introduce an oncogene and disrupt a tumor suppressor gene in a brain organoid, causing tumors to form. She did not use tumor transplantation. Instead, this researcher induced tumors to form within a brain organoid.

Going in the same direction but using another technology, a research team is set to make cerebral organoids from a person's own stem cells. Personalized organoids may solve immune system rejection among several issues. Would it work on both genders equally? Some research scientists are skeptical about meddling with the intricacies of the brain. They claim we do not know enough brain physiology, signaling, electrical, and chemical systems. The discovery of cerebral organoids was done by Madeline Lancaster, a cell biologist, at the Institute of Molecular Biotechnology of the Austrian Academy of Sciences.

I cannot help but wonder why CRISPR researchers have not gone after cancerous cells. There are several types of cancers. There are two types of bacterium found in the gut that might boost the risk of colon cells turning cancerous. Our intestines are host to trillions of bacteria. A very thick layer of mucus separates the surface of the colon from our uninvited tenant, bacteria, billions upon billions of bacteria. Making it more complicated, some people develop small growth of tissue from the surface of the mucus membrane of our intestines called polyps. In people with genetic propensity, some of these polyps were found with patches of mucus that were invaded by bacteria, mainly, *E. coli* and *Bacteroides fragilis* each carrying a gene that encodes cancer-promoting toxins. [67]

---

67. *Nature* 554 (February 8, 2018): 149.

Furthermore, a recent study found that both innate and adaptive lymphocytes shape microbiota in the intestines. It was suggested in studies in mice that colon cancer treatment could be improved by simultaneously targeting two signaling system in tumor cells. David Horst at the Ludwig Maximilian University in Munich, Germany, and his colleagues analyzed 328 human colorectal tumors and found they tend to have high levels of Notch activity at their centers. Signaling among cells is crucial in any investigation. During this investigation with mice, signaling coming from the center and those coming from the edges was different. Targeting both proteins slowed colon cancer growth better than targeting either alone. We must remember that our intestines are host to billions upon billions of bacteria. Consequently, our gut is a place where many of our diseases originate.

There are billions of chemical processes taking place right inside our belly intestines. Most bacteria in our intestines are not a threat to our health; there has been some type of mutual adaptation and accommodation for them. There is no threat to my life if a balance is kept between good and bad bacteria. It is in the best interest of our health for us to avoid disrupting the balance of good and bad bacteria in our intestines. Try not to bring into your body more life-threatening pathogens than your immune system has successfully overcome. Our immune system produces T-cells that travel in our lymphatic system looking for invaders and enter battle at any moment against an enemy they consider life threatening.

You can add strength to your immune system. There are several ways we can do that. First, wash your hands with soap and water before eating. Going in and out of bathrooms, we must be very careful when touching the doorknobs. There are a lot of *E. coli* bacteria anywhere in those rooms despite cleaning. Keep a good distance from people coughing a lot. Saliva and mucus from sneezing are carriers of multiple microbes. When you go to a hospital or make a doctor's visit, you will be surrounded by sick people. Be careful. You do not become contaminated. Prevention is the best medicine you can ever have.

We have several diseases for which there is no cure at the present time. The therapeutic tools that are available at present are basically palliative, meaning short-lived remedies. Among this list is cancer in its many forms. Except for surgery, the remaining therapeutic tools are aimed at relieving pain and suffering. However, cooperation among cancer researchers worldwide give hope in finding a cure within the near future.

Cells that help create bones also promote lung tumors in mice, reported Mikael Pittet of the Massachusetts General Hospital in Boston. A mouse with lung cancer has high number of bone-forming cells called osteoblasts, which are found in the bone marrow. Reducing levels of a subset of osteoblasts known as Ocn+ cells in mice led to slower growth of lung tumor. These cells were traced to white blood cells called neutrophils. Neutrophils can invade lung tumors. People with various cancers who have high neutrophils levels do poorly.[68]

This seemingly unimportant discovery would promote further interest in cancer research when shared by researchers in worldwide meetings and conferences. Worldwide research cooperation making use of communication tools available to members could offer opportunity to find the best place to fulfill their academic and research ambitions. While many PhD graduates look to academia to fulfill their careers, others go to the industry, tired of writing papers for scientific journals and grants. Some of the reasons for going to the industry seem funny but practical and realistic. During several surveys, I have come across these comments: "It is an opportunity to make drugs and stop pain instead of writing papers."

"I will be helping people with new treatment discoveries. It is the satisfaction of seeing people recovering from what was considered an incurable disease, enjoying themselves at home or at work."

In my case, I was very happy to be a clinician: practicing, supervising, and teaching small groups.

There are outstanding professors in academia who miss the experience of drug making and discoveries. They claim that on the halls of universities, it is not easy to see the fruits of their scientific expertise and labor. Often you are trapped and do not know it. The

---

68. *Science* 358 (2017).

pressure of writing papers, prospect of promotion, family demands, and needs do not make an easy transition from academia to the industry. During my search, I found a very interesting example of what I believe is what is going on satisfaction and career demands. A prominent and professionally successful immunologist, Shannon Turley, at one time said, "I am an academic at heart. I did not see myself in industry at all. However, after a visit to Genetech's laboratories she felt she would be able to continue her fundamental research while helping to transform discoveries into therapies for patients."

Here is my conclusion after several successful jobs in the health field. We grow physically and mentally, making important decisions at crucial stages of our life. I was exposed to stimulating external triggers that excited several neuronal connections that altered my behavior positively. As my brain changed physically at synaptic level, I became aware of my needs at the cognitive level. I made corresponding job adjustments and changes.

I have concentrated my attention on the genesis of my primordial memory in the brain. However, as I proceeded with my investigation, I must have asked the following questions: Are all my memories made, formed, and stored in the brain? How did my body remember past self-defense tools to protect itself from invaders and predators? There were many questions coming up in my mind that almost put a stop to my investigation. Therefore, I chose to study just the most urgent one for me. What does the brain consist of? How is the brain related to the rest of my body? Are my toes part of my brain? For example, I feel pain when I am walking barefoot and I step on a sharp object. Why is it that I lift up my foot after feeling pain and not before? What is the brain composed of to be able to store a memory? What is the structure of a memory? Are memories stored as pictures of objects? Most people would consider the following question naïve, but is it? Where is the brain located in your body?

Chapter XX

The Brain's Main Components

The brain is composed of a huge number of neurons counted in billions. It consists of the central nervous system located inside your skull and the medulla extending down your spinal cord. There is a second system called the autonomous nervous system, which is divided into the sympathetic and parasympathetic nervous systems. The sympathetic system places you in action while the parasympathetic brings you back to normal action and behavior. These brain systems communicate through cells known as neurons. Neurons are composed of a central body with many dendrites receiving signals from other neurons. Besides short dendrites extending from the cell body, there is another long fiber in the shape of a tube known as an axon. The axon delivers information to another neuron. Dendrites receive information while the axon delivers it. The message comes from a presynaptic neuron to a postsynaptic neuron. The message is a chemical message, but it is delivered by electricity initiated in the neuron's body. The chemical message is delivered at a synapse (space between two neurons) to an expecting or receptive molecule of a dendrite. Synapses can be formed at the cell's body too. The electricity originating in the cell's body travels inside the axon, which is insulated by myelin just as an electrical wire is with plastic. The myelin insulation prevents electricity from spreading to nearby neurons besides facilitating speed. The chemical message is also facilitated by potassium, sodium, and calcium pumps almost at the end of each axon. Some of the chemical messengers crossing a synapse are the following: dopamine serotonin, acetylcholine, glutamate, norepinephrine, and GABA. Glutamate is the strongest excitatory neuro transmitter while GABA is an inhibitor.

Most of brain cells do not code for protein synthesis. These nonprotein coding brain cells are called glia cells. However, some glia cells, such as microglia, perform a very important role in my body defense system. It is the immune system of the brain and protects it from virus and bacteria. Actually, microglia does its work from our lymphatic system.

Curiously, if glia cells do not code for the synthesis of proteins, where are proteins synthesized? The whole body is made of proteins. We can distinguish the shape of many organelles when we look inside an animal cell. And shape determines functions. Peter, my ever-present friend, reminded me that we need a microscope to be able to see and appreciate a living cell. There are several organelles traveling inside the cell plasma. Each one has a specific and specialized role to perform. We will find tiny ribosomes at the opposite side of the nucleus within the cell membrane. Ribosomes are assembly stations for protein synthesis. Without being repetitious, proteins come in different shapes.

From the nucleus of the cell where DNA is located, a single-strand molecule named messenger RNA comes out, and its destiny is a ribosome. This messenger RNA is the template or blueprint for protein synthesis. Another single strand named transfer RNA will assist the messenger RNA in its travel to a ribosome. This complementary t-RNA and messenger RNA are assembled together during the process of protein synthesis. Transfer RNA comes from the nucleolus located within the cell's membrane. The nucleolus has its own membrane. There are several RNA strands with specific roles. All RNA come as single strands while DNA is double stranded. The two separate strands are wound around each other in a helical formation. This is the famous double helix James D. Watson presented in Long Island, New York, in 1953. There began the century of DNA and RNA double helix. Messenger RNA most always code for a single protein. The four bases in DNA are adenine, thymine, cytosine, and guanine. In RNA, uracil replaces thymine. Both strands are held together by hydrogen bonds by positive and negative atom's charges. RNA molecules are very versatile. CRISPR-Cas9 has an RNA component.

Let's continue our discussion of neurons, axons, synapses, memory, and electricity. Axons may have sections uncovered by myelin to add more electrical speed down the axon's terminals. Axons may be tiny as in interneurons while the longest one stretches from the end of the spinal cord to your toes. It is an awesome sight to see billions of neurons and axons communicating among themselves in modern scanners. The inside of a neuron cell is set at specific voltage when it is at -70. This is considered a resting phase. However, when excitatory impulses come to the cell body, it can overimpose the resting balance and go as high as -35 millivolts, and the cell fires. Consequently, an excitatory message travels down the atom. Glutamate is the strongest excitatory neurotransmitter. GABA, (gamma-aminobutyric acid) is an inhibitor. Firing in the cell is not just a matter of numbers but also a location of synaptic connections on the cell body and dendrites.

As I noted earlier, the inside of the cell is a wonder world of vesicles traveling in several directions. The chemical message they carry must be protected from being cleaned up by enzymes and lysosomes. It may seem that there is some type of disorder within and among cells. There is not; there is order. There are corrective interventions by the cell itself. There are corrective enzymes besides internal regulators. According to biochemists, most mistakes are provoked during protein synthesis. It occurs during the transportation process with RNA strands. It may show you that code translation is not an easy job. RNA strands are engaged in carrying molecules to ribosomes and engaged in amino acids. There is hardly any process within my body that RNA strands are not involved.

My body must have stored memory for a long, long time to be able to survive thousands of years and be able to tell the story. My two basic life acids, DNA and RNA, have made it possible. It has been assisted by a few allies such as the immune system, mitochondria, and energy-producing molecules. Not a small job.

All work done by the RNA is known as translation while transcription is done by the DNA. Messenger RNA comes out of the cell nucleus while the nucleolus is the home for RNA single strands. In the double-strand DNA, each base pair is held together by simultaneous attraction. There is a positive hydrogen atom to two other atoms with a negative charge. Positive- and negative-charged atoms attract each other. These DNA letters, *A*, *T*, *C*, *G* with their backbones known as nucleotides, shape the famous spiral ladder of the double helix. This is my genetic code packed together in a linear fashion in just four letters. Awesome, my life history written in just four letters!

Chapter XXI

Genes

My life, my genes, my memory are kept together and alive, encoded in four letters for thousands of years. Now, I can travel or dream about visiting places and countries where my ancestors of long ago lived and contributed to our civilization. Of course, I would like to visit cities and towns along the Euphrates and Tiger rivers: Babylon, Nineveh, Persia, Troy, Athens, and ancient Jerusalem. It has not escaped my curiosity that my ancestors of many generations ago went up the Caucasus Mountains and fished in the Caspian and the Black seas. Dream, yes, dream, neurons, axons, and memory, but it is all written in my genome. This planet is mine to live and enjoy. My dream about my ancestors of thirty or sixty thousand years ago is a relatively recent dream. A team of investigators, including Swante Paabo from the Max Planck Institute in Germany, reported that Neanderthals and Denisovans are extinct groups of hominids that separated from each other more than 390,000 years ago (*Nature*, Vol. 561, September 6, 2018, p. 113–116).

At the same cave at Altai Mountains in Siberia, Russia, S. Paabo and his team of investigators found genes extracted from teeth and fractions of bones from Denisovans and Neanderthals of around 432,000 years ago. Therefore, my ancestors' genes are recent history. According to some geneticists, modern humans replaced both Neanderthals and Denisovans. There is no evidence, at present time, that neither

Denisovan nor Neanderthal ever achieved a cultural or civilization level comparable to modern humans. Actually, modern humans have achieved more progress during the last 200 years than our ancestor did in 200,000 years. Alphabets, writing, and books are very recent tools to move forward and succeed.

Genes are segments of a DNA molecule composed of linear sequences of nucleotides that repeat over and over again. Each one nucleotide consists of a phosphate group, a sugar, and nitrogen-containing base. The DNA molecule has four bases: adenine that pairs with thymine, guanine that pairs with cytosine by positive and negative charge. The DNA molecule determines the order of amino acids that will go into all proteins that build your body and mine. There are three nucleotides for each amino acid. The DNA molecule is in the nucleus of a cell and produces the molecule messenger RNA, which is the template for protein synthesis. If messenger RNA brings out/ reads the letters *CCU*, the amino acid proline will join this molecule. However, if messenger RNA reads *CUC* instead of *CCU*, the amino acid leucine will join instead. The RNA molecule has uracil taking the place of thymine. During the sequencing a genome, if you know the mother molecule, you will easily recognize the daughter molecule because they run parallel to each other but in opposite directions. If the mother molecule begins with *A*—adenine—a daughter molecule must begin with *T* (thymine) and so forth.

Some biologist calls it antiparallel to each other. Genes are made of DNA. Proteins are made from amino acids, and genes nucleotides. How many nucleotides are there in a gene? It depends on each gene, and it could be counted by the thousands. How many amino acids are there in RNA? The answer is none. RNA is a basic acid. What is an amino acid? Amino acids are organic, which means they have life. These amino acids are compounds containing amino- and carboxyl-functional groups. There is also a side chain to the amino group. Amino acids are formed from different combinations of carbon, hydrogen, nitrogen, and oxygen. The wonder template for proteins synthesis is the messenger RNA that comes out from the cell nucleus of the macromolecule DNA. I wish I could celebrate the day when DNA,

RNA, and mitochondria molecules joined to form one macromolecule, our DNA. This elementary introduction to molecular biology is very easy and easy. Just remember to practice some basic rules, and you will be master of your life and destiny.

I like to repeat some of the things mentioned earlier. Amino acids are small chemical compounds that contain both an amino functional group and a carboxyl-functional group. During protein synthesis, both molecules are transferred to ribosomes where they are translated into peptides chains. There are peptide chains of several lengths, including proteins. Another way of saying it is that amino acids are organic molecules that consist of a basic amino acid group (-NH_2), an acid carboxyl group (COOH), and an organic *R* group/side chain, which gives amino acids its characteristics. Both, messenger RNA and transfer RNA must go to ribosomes for protein formation. I cannot help it, but it is a great memory. I will add that a restriction enzyme has another great memory. Restriction enzymes must remember where to cut the genome to begin the process of protein synthesis. Everything in my body seems to be based on memory. Lifting my feet to walk appears to me to be another great memory.

I repeat here that amino acids are primarily composed of carbon, hydrogen, oxygen, and nitrogen. There are many amino acids, but only twenty are used by my body and yours. How these great molecules, DNA, RNA, and mitochondria acids came together to form life and encode and store my life history is another wonderful memory. This life memory is stored in my brain, making use of neurons, axons, synapses, chemical compounds, and electricity. What would man be if there were no electricity? It runs my life, and I cannot see it.

Taking a few steps backward, I would like you to place yourself mentally and physically enjoying a past adolescent experience. Your cooperation is greatly appreciated to be successful. Please do not disappoint me. We have visual memories, auditory memory, taste memory, touching (skin) memory, and smell memory. Join me in this hypothetical illustration. A boy and a girl walked down the river and had the first intense kiss of love under a tree full of flowers. Six months later, they returned to the same tree to celebrate their love to each other.

However, their meeting ends up in bitter accusations of betrayal and dishonesty. Would this memory be formed and stored on the same cell or cells (neurons) as the first-kiss memory? To be more precise, would it be the same pre- and postsynaptic cell? There must have been a lot of anger, verbal threats, and perhaps tears. In an anger situation, glutamate, an excitatory neurotransmitter, seems to be always involved. I believe the first-kiss memory between the boy and girl is formed somewhere in groups of cells in the limbic system. This is a system of joy, pleasure, good time, and happiness. A problem is that the kiss on the mouth seems to have spread to other parts of the body. The neurotransmitter dopamine is king in the limbic region of the brain. These two brain chemicals do not seem to be friendly to each other. There seems to be at least two different memories: one for love and another for anger and perhaps hate. This is the same boy and the same girl under the same tree along the river. The purpose was to celebrate their commitment to each other. Can you help find out the possible locations or location for this seemingly antagonistic memory? Have you thought of the emotional depth and intensity involved with both memories?

During the first kiss, the memory may be limited to the limbic system. The second memory might have begun in the limbic system, too. A scanner such as an fMRI would be helpful. It would show, among other things, blood flow and lights within each region containing sections of the memory. Someone said that love and hate live together, and that one does not exist without the other. You cannot hate anything that you have not loved. It seems that love and hate memories are destined to share the same roof, the limbic lobe. Is it reproduction or is it the limbic system that joins together a man and a woman?

What are neurons and axons without a synapse? Not very much. For sure, you and I would not be here. The junction of two neurons is called a synapse. The presynaptic neuron fires an electrical impulse after reaching an action potential. Actually, the impulse begins at one of the brain cell regions named hillock. After leaving the hillock, the electrical impulse will travel through an axon to deliver a chemical message. An axon is the longest extension from the neuron's body. During its way to the next neuron, the electrical impulse in the axon will be assisted by three chemical pumps to deliver its chemical message. The three pumps are potassium pump, sodium pump, and calcium pump. The terminal end of an axon breaks into separate and tiny branches, which interface with dendrites of the postsynaptic neuron. A dendrite in a synapse is a receiver or entrance point of information to the neuron. An axon, which is the delivering section of a neuron, can be more than a meter long. It carries out a message, a chemical message from the neuron's body. A neuron has four basic sections: axon, dendrite, hillock, and body. They are distinct regions, each different and specific, carrying out separate and specialized functions. The neuron's body has many short extensions called dendrites. These dendrites can be counted into thousands.

The cell body contains the nucleus. It is the site in the cell where all neuronal proteins and most membranes are processed. Some proteins are formed in dendrites, but no proteins are made in axons and/or axon's terminals. Ribosomes are needed for protein expression. Despite its length and size, axons do not have ribosomes. The cell body also assembles vesicles transporting chemical materials down the length of the axon to the terminals. These microtubules can also move up the axon toward the cell body. Almost every neuron has a single axon, which is specialized for the conduction of a particular electrical impulse called action potential. An action potential is a series of multiple, unpredicted, and sudden changes in the voltage across the plasma membrane. When a neuron is in the resting state, the electrical potential across the axon membrane is around -60 mV; it is the inside negative relative to the outside. At the peak of an action potential, meaning stimulated, the membrane potential can be as high as +50 mV inside positive.

It is safe to say that my body is a chemical organism that moves around and functions based on brain cells that generate electrical impulses. These brain cells—neurons—are the basic functional units of our nervous system. The central nervous system consists of the brain and the spinal cord. There is, besides our central nervous system, the peripheral nervous system that consists of neurons outside the central nervous system. It includes sensory neurons and motor neurons. Sensory neurons bring signals to the central nervous system while motor neurons carry signals out the central nervous system. For instance, I see a big dog coming in my direction. My sensory neurons for vision will contact or signal the central nervous system. Instantaneously, motor neurons in the neocortex will place me in a state of action. If there is imminent danger, the amygdalae can override decision-making stations and provoke the hypothalamus to release emergency hormones such as adrenaline.

The bodies of some peripheral nervous system are located in the central nervous system. They are known as motor cortex located all way up in the neocortex. These hyper cells have long extensions. They can use innervate neurons to reach your toes. In addition, cell bodies of the peripheral nervous system, sensory neurons that provide signals to touch, bodily positions, pain, and temperature, are located outside the central nervous system. They are called ganglia. Sensory information from my surroundings will end up at my central nervous system for processing. In general, I can add that the peripheral nerve system consists of nerve fibers that branch off from the spinal cord and project to all parts of the body. These fiber projections will include the neck, arms, legs, torso, skeletal muscles, and internal organs. Therefore, the peripheral nerve system is a division of the human nervous system containing_*all* the nerves that are located outside of the central nervous system.

A decision will be made at the neocortex and respond accordingly. The frontal cortex is a very important decision-making region of our brain. Of course, age is a determining factor. My body is eighty-nine years old and cannot move with the speed and agility of a youngster at age twenty. Motor neurons receive signals from other neurons and convey commands to my body's muscles, organs, and glands. For instance, during the summer, I like to walk barefoot in the backyard of my house. Occasionally, a sharp object may get on my way and cause pain, and blood maybe spilled. Why? It happens when I do not carry a cane with me. I cannot stand on one leg. I have Parkinson's disease and lose my balance relatively easy and fall to the ground frequently. By the time I can move my foot, it is too late. The sharp object has pierced my foot membrane, and blood comes out.

I said earlier the longest axons stretch out from the back end of my spinal to my toes. The bundle of axons inside mine and your spinal cord is called the medulla. It is a central highway for information coming up and down your body. A fracture to your spinal cord involving the medulla can leave you paralyzed for life. There is a lot of memory involved in the human nervous system. There are some brain diseases connected to nonfunctioning brain nerve cells.

Some bio-scientists believe that protein synthesis, synapses, and RNA translations are the source of several brain diseases and mental disorders.

Love thyself so you can love your neighbor. Do not hold a grudge against anyone; it only leads to unhappiness.